Encyclopedia of Technology and Science

Volume 111

The science

Prof. Dr. Sami AL-Mudhaffar

Contents in brief

Preface

Academic disciplines have evolved with the development of sciences and various new disciplines such as engineering, agriculture, science and total treatments that began in the nineteenth century, whereas the twentieth century indicates other disciplines such as business   management, journalism, information and library science, economics, politics and world affairs were added.   Each state has its own special methods to determine its own disciplinary university and identification numbers, graduate students and the quality.

The world witnessed in the twentieth century breakthrough in all fields and scientific trends, so there are no boundaries between different disciplines. For example, medical science requires engineering science and recent tests of modern science depends on the physical, chemical extraction and analysis and also relies on mathematics to lay the groundwork mathematics.

The progress and development in pure science, for example, develop new subjects and disciplines and specialties of science. New interfaces were not known during the first half of the last century. The results of these major changes in curriculum and build up research transformed these developments to the university curricula. Seminars and researches are now carried out in different ways including:

- Bachelor based on the study and theses.

- Some universities in Britain and Germany developed curricula at the level of initial studies that include research and study.
- Dividing the present fields such as industrial chemistry, chemistry of life with medical side and other disciplines in the branches of pure science.
- Developments of competencies.

These terms of reference have been developed in American universities in physics, chemistry and mathematics, to prepare graduate in some sectors such as engineering, chemistry physics and chemistry, agricultural engineering, and medical studies.

- Adding assistance topics.

Some topics have been added as assistance of many of the terms of reference of pure sciences, including education, literature and library services and use modern machinery and computers.

- Other disciplines (Sandwich)

In about two hundred years, there will be four possible scenarios for future reference, namely they are:
- Very pessimistic.
- Cautiously pessimistic.
- Cautiously optimistic.
- Eager for growth and technology.

In future new generations will be created and involved, the so-called future shock goes through analysis that includes:
- The future shock is a severe illness that affected increasing numbers of human beings and can be called satisfactory inability to adopt to rapid change.
- Reactions to the future depends on what is known about the ability to adapt recalled hypothesis in the 1st. instance, that the proliferation of psychiatric and neurological a lot of people due to the shock of the future and technological progress and scientific facts have not been able to create awareness on assimilations

Basically, the science is understanding how things work and why, and gave us to survive, and improve our lifestyle in the process, that science is the most important element of our existence. The recognition of the importance of science is historically known through different civilizations (Pharaonic, Sumerian and others), passing through the efforts of Muhammad Ali in Egypt and study of Shibley Achammal (1853-1917) of the theory of evolution and cell biology which he analyzed the reasons for the weakness of the Ottoman Empire.

Science is a cultural activity achieves the objectives of economic and political community, and according to that science and technology are major forces support the historical, social, national and international change, vary from one community to another, and the internal and external challenges vary from culture to culture.

The Middle East, which intersect three continents of Asia, Africa and Europe it represents the advantages and disadvantages of his power. The beginning of science started with human himself, where the scientific knowledge passed on from generation to generation through the professions, and through the end of the nineteenth century it was dominated by the religious character of scientific knowledge and focused in the church.

It has been focusing on enriching the life with scientific research and according role of science as social power and after that science curricula evolved and reform movement was appeared for the need for the development of curricula for scientific creativity, and in the year (1983) a report was issued entitled "A Nation at Risk "in the United States, and it becomes a warning signal to the community.

Science education should contain the principles and conditions, including early learning of child and in primary, preparatory stage, to develop the ability to use the scientific method to solve problems and to gain scientific knowledge and have a role in achieving the acquisition of facts by revolution and the acquisition of the concepts of scientific principles..

to the fact that scientific knowledge intended to develop the capabilities and skills of individuals in order to meet the requirements of the various aspects of life, as well as the community. Accordingly, the development and developing the capacity to build the skills of human individuals.

One of the most important things science gave us security – we figured out how to take care of our physiological needs, as well as our physical needs. We made the most out of science by initially trying to understand how things worked, and then making them work to our advantage; the result is that, besides offering us what we needed, science allowed us to create what we wanted.

Science allowed us to understand how things work, but also how to make things happen. Social sciences made us understand that, if we want to survive, we need to work together in an organized matter.

That science is a cultural activity to achieve the goals of economic and political community and according to that sciences are major forces support of the historical, social, national and international change, vary from one community to another, and the internal and external challenges vary from culture to culture, in three continents of Asia, Africa and Europe, as it represents this site and the dominion advantages and disadvantages.

During the mid-twentieth century, it has been focusing on enriching the life scientific research and according to the needs of society and the role of science and social power and evolved its impact science curricula emerged reform movement in the seventies calls for the need for the development of curricula for scientific creativity, In (1983) issued a report entitled A Nation at Risk in the United States, a warning signal to the community.

Due to the fact that scientific knowledge intended to develop the capabilities and skills of individuals to cope with the requirements of various different aspects of life, as well as the community. Accordingly, the development of skills and capacity development of individuals is to build a human.

Man plays a distinct role in the knowledge that the first human activity and the second humanitarian component of social, scientific and general structure inherent rights throughout his life. For the life of human being is a living organism has a specific building my body interdependent with the psychological condition and functional.

Despite these perceptions Man builds different perceptions in the general philosophies when it is the ideal object of philosophy spiritual exercises will and is responsible for his actions. Mentions Plato that man is composed of two parts, one of which belongs to the world of ideals (self) and the other for the world of sense (the body) and stressed the idea of bilateral mentioned, while the realists believe that man is an organic entity, social and natural philosophy and tendencies suggest that the human soul's finest and present continued evolution of the future and education is necessary as long as they continue to longevity and Islamic philosophy thread dimensions of human nature (body, mind, spirit, heart) and consistency and balance.

Scientific knowledge and life there are several theories as to why the emergence of life some of them is the fall of organic molecules on Earth from comets and cultivated land life of intelligent beings in advanced planets, there are two trends are clear in the study of the mystery of the creation of the universe and life, the first thought that there is a very in it and secondly, denies the existence of objective but to The presence of the shell, and subjects of interest in the mystery of the creation of the universe and of life, the origin of the universe and the evolution of different theories that talk about its inception .

Amazing developments have taken place in the chemical sciences particularly during the second half of the century, including implicit and other interfaces. Developments on the implicit content and the vocabulary and mechanisms are known in chemistry and provide improved or new interpretation of events and phenomena and chemical reactions, as a result also of new subjects and disciplines within the science of chemistry itself. These developments have led to the opening of new channels in scientific research and technological innovations such as chemical industries to create new chemicals, or chemical industries, and new techniques.

The developments of the second type of chemical sciences interface had addressed the disciplines of science linking chemical sciences and applied various treatments. These developments have led to the developments of science or the new terms of reference were not known before.

The second half of the nineteenth century witnessed a series of discoveries of life such as, a serious (cell) theory at the hands by Matthias in the plant and then Theodor ishvan in the animal. Since of plant and animal are composed of cells, that evolution (cell theory) and their development is considering critical stage in the progress of life science similar to atomic energy The human Physiology science as one of the branches of the life science that refers to the amazing facts explaining the greatness of the Creator and accuracy of the details and secrets. The digestive system for example (the greatest chemical plant in the world) including, the methods of food analysis of chemical analysis of various surprising and distribution of fairly Safe food distributed to millions of living cells. In view of these living cells the issue of causal efficacy and secret of life that fills justify the astonishment and admiration for self-cell, while adapting to the requirements of their position and circumstances.

If we explore the science of life, we will find another secret of that biggest secrets, the secret of the mysterious life, which fills the moral conscience of mankind, with the concept of divine fear and faith, firmly established in it. The theory of self-regeneration was collapsed at the depth but the unequivocal scientific experiments, demonstrated the invalidity of the theory of self-regeneration.

The material basis of life science was examined and then basically spread the idea of elements. The atoms are spread better for the basic materials of the universe and second nature that the elements consist of a central core electrons of the nucleus orbit (negative) and the nucleus contains protons and neutrons. Attempts were made to alter the material to absolute energy, no electric charge. In other words removing character from element in the light of the theory of relativity of Anstein, where the body mass is relative, not fixed, and increase with the speed according to Anstein equation energy = mass of × square of the speed of light and mass = energy ÷ square of the speed of light. As a result, the atom, including of protons and electrons are condensed energy. Appeared in various forms and multiple images, whereas materials has been converted into energy and energy to the material.

It follows from the views put forward that the original materials the world-life the reality show one common in various forms, and the physical properties of compounds are accidental such as the liquidity of water is incidental, not self-evidence, since it is consisted of two toms and possible separation these two elements from each other and the status of water disappear completely.

characteristics of the simple elements themselves are not self-rule but are incidental to the material. That such material characteristics become the light of the above facts incidental, it is encroached to be among the identified energy and philosophically, the presumption of material in the world of life on the top reason capable for denial, as well as of effectiveness.

Living organism and the secret of life

When studying living organisms as secret of life and mysteries dependence of the of life it might be important to consider many concepts such as the idealism that reflect the objects of which they are the realities of life and is found independently of sense perception, and is the way of our thinking for our perception. As a matter of realism philosophy, another question is the position of living objects reason for all the phenomena of existence and the universe, or bypassed to another reason represented the deepest area of the material, another beyond the spiritual and the last living as a reason to click the Spirit a realistic concept of Divine (the divine), and this concept does not mean dispensing with the reasons natural or something to rebel against the facts of sound science, a notion that God is deeper reason for science to explore the wider area, including the continuing nature of living organisms.

The Philosopher, whether materialistic or divinely believes the positive side of science, such as exploring the unity of life of organisms with the general knowledge that is not in issue, the scientific issues , divine philosophy materialistic philosopher material. Agasiz introduced in the year (1858) the idea represented by all kinds of organisms created by special acts of erecting force. This opinion is consistent with view of both Al-Razi and Pasteur and settled their minds with that every living being that must be generated by an organism like him. Also, Hermann Erhard Brichter has stated that each living thing is eternal and produce only from the cell.

The cell as secret of life

The human digestive system which is an arm of the human physiology of man, explain grandeur of the Creator and the accuracy in detail of the multiple and various secrets. The digestive system is sophisticated chemical laboratory having different methods of food analysis then the food will be distributed equitably to millions of cells that make up the human body consideration involving the secret of life and admiration for the cell. These cells are different technologies for tissue engineering and in the digestive system nearly two hundred thousand reaction within 24 hours.

Some of which are the heart muscle shrinks and flattens millions of times during the whole year tirelessly, to obtain the necessary energy for thinking and movement and speech, including also the disposal of waste and toxins in the body, looking at the cell that its approach is one of the secrets of life. These secrets are adopted according to serious lay cell; it is one of the discoveries, theory at the hands of Haydn and Schwan then cell theory, considered a critical stage in the progress of life science, similar to atomic theory in chemistry.

The cell in the body of an organism is also similar to personnel in the communities or the living cell act as technique specific to perform a particular job. The cell becomes a plant or accurate chemical plant. The nerve cell act as a system, for example electrochemical transformation of chemical energy to electrical energy and electrical energy to mechanical energy or kinetic energy. There are also some cells manufacture of the of hormones and other life products used as system used defensive attack their products all exotic and cells in the process of purification and filtration and the cells that absorbs.

Furthermore arises from a single fertilized cell, tissues are various heterogeneous tissues, the different organs and different functions, the bones, muscles, cartilage and twigs, leather, and the blood vessels. Then a living cell had made specific technologies in particular, for example, nerve cell is electrochemical system that can shift the chemical energy into electrical energy and the last to mechanical or dynamic or may become a cell laboratory or chemical plant carried hundreds of chemical processes complex and there are cells specialized in the manufacture of hormones and the other to produce biological weapons of and cells of the nomination and purification.

It is clear that the tissues that originate from a single fertilized cell then divided into thousands of cells to materialize into the bones and muscles, cartilage, twigs, leather, and the blood vessels, these tissues are formed in the early embryos and mutate into organs and systems in a stand- alone, but integrated in the performance of its functions.

To the information of originated from the cell that there is strange power lies in living cells. Walker, professor of Plant Physiology say (that components of a cell arranged in a strange way in which life emerged). But researchers still are unable to make blood cell components and accurate knowledge of this so-called the mystery of life.

Nucleus and secret of life

There are at the center of the cell mass of material in the spherical or oval clusters of objects in the body and continues to represent the mysteries of life and plans, regulations, and ideas of life.

In the cell of the human body forty-six chromosomes except the egg (cell reproduction female) sperm and egg, each containing 23 chromosomes (half the number in the human cell non-reproductive). The secret of life is due to this process of somatic cell fusion where each chromosome separated into two parts, it becomes in each of the cells of the fission 46 chromosomes. The chromosome (multi-genes), each genetic factor arranged in two strands one received from the mother and the other from the father.

The nucleus synthesize nucleic acids, seen in the central nucleus filamentous structures and spread over its surface of granules of quick-impact dyes include the nuclear network and the network should be clear when they are not in the case of splitting and dividing at smaller and thickening of these lines is called chromosome. The chromosomes in a cell division looks like similar pairs of fixed shape and fixed number for one type of living organisms.

(Russell and Wallace) says that the cell nucleus is not chemical but structure if analyzed and during the processes of analysis of the most secret mysteries of life may be lost.

Molecules of life, which are building the organism

The space of cell is containing the liquid water containing the various ions and compounds with molecular weights of small, medium and macroscopic, and it is possible to measure the ion composition in each cellular organelle, where each one of them has different ionic compositions.

The sodium ion "Na +" ion is the main ion out the cell in which the 140 mM / L is also found a positive ion in the fluid cell in the Interior position that the , is potassium "K +" Cation cell procedure. There is 46

magnesium ion "2 + Mg" in all cellular spaces inside and outside but with lower concentrations of sodium, potassium and chloride is "CL-" ,the main negative ion outside the cell, with hydrogen carbonate ions "and" small amounts of phosphate and sulfate, and the proteins carry a negative charge at pH 7,4 in the tissue fluids.

All living cells contain a different chemical components of water 70-90% and 2-5% of inorganic ions such as sodium, potassium, chloride and sulfate, and magnesium ,carbonate molecules of life, as well as small, medium and macroscopic molecule that constitute 8-25%.

It has been proven that all the elements in the periodic table of Mendeleev's constitute in the composition of the organism divided into small elements and large. The carbon, oxygen, hydrogen and nitrogen constitute 96% of the elements in the cell, while calcium and phosphorus constitute 3% and each of potassium and sulfur, iron, sodium and chlorine 1% There are very small quantities of the elements iodine, magnesium, copper, manganese, cobalt, boron, zinc, fluorine, selenium and molydenom.

The chemical side is concentrated in the molecules of life on the carbon, which constitutes about 50% by weight bio- molecules are characterized by life-covalent bonds, four of which related to carbon stubs and have different angles of particular value from one carbon atom to another in different molecules of life and because of that there are different types of building structures with three-dimensional, these structures contribute to clarify the complexity of the cellular composition with particular reference to its failure, as well as various forms. In addition organic compounds are characterized by free rotation.

The tetrahedral bonds emphasizes on individual carbon atom of the very important properties of organic molecules and the presence of four different groups or different atoms connected to carbon and the last become non-symmetrical (a carbon-atom covalently bonded with four different groups) and composed (which every one of which is mirror images of each other) with a symmetric arrangement in space and called isomers mirror of light for the chemical similarity of the interactions but differ in physical properties of the rotation of polarized light.

Using the United States as an example, some of the topics to be discussed are the views of public officials who influence the distribution of research funds, the response of funding agencies and the views of scientists.

Finally, we shall look at the co-evolution of science and society and attempt to draw some conclusions concerning their related future and the implications for the future of technology.

Public officials who are involved in setting or influencing science policy have expressed opinions that indicate that they intend to change the basis for supporting research and development.. the public officials wish to alter somewhat the pattern of funding for science.

Their motivation is to orient research more toward programs that, for example, ensure a stronger economy and improvements in the environment. It is becoming increasingly apparent that those public officials who control public funds, will be reluctant to fund research programs that they consider unrelated to national needs.

Academic disciplines have evolved with the development of sciences and various new disciplines such as engineering, agriculture, science whereas the twentieth century indicates other disciplines such as business management, journalism, information and library science, economics, politics and world affairs were added.

The world witnessed in the twentieth century breakthrough in all fields and scientific trends, so there are no boundaries between different disciplines. For example, medical science requires engineering science and recent tests of modern science depends on the physical, chemical extraction and analysis and also relies on mathematics to lay the groundwork

Chapter One

General Aspects of science

Science and human being

The event in 1996 was exciting, the birth of Dolly in a somatic cell into a specialized egg-enriched after removing core and planting it in the womb; the most important point of this event is the return of specialized cell and embryonic stable situation after losing this status. The other development in science is the production of sufficient quantities of food in the world. Many thinkers expect that the world will see a lot of problems, related to scarcity of resources and energy, such as increased pollution and population explosion. Most studies of future ending 2025 required further means such as:

- Technical means (computers to store and recall information).
- The use of special programs to make the prediction.
- Many experts, technicians and programmers.

Possible divisions of future studies in science include three types; this division is used haphazard to simplify them, as it is difficult to separate the three types of studies, from each other, such as:

- Studies that rely on prediction (what will happen in certain area of science).
- The overall outlook studies that rely on intuition which looks at the impact of current scientific achievements on the future of humanity.
- The overall outlook studies relied on detailed statistical information within the mathematicians programs on computer models.

Future Studied are focusing on several scientific areas such as renewable energy, genetic engineering, biotechnology, electronic industries, the manufacture of computers and communications, telecommunications space and material science. Researchers expect, the use of satellites for the transfer of solar energy to micro-wave stations can broadcast to the ground as the receipt and then transformed again into energy that can be used, then in the field of genetic engineering, many recalled of the perceptions of future scientists in the following futurism:

- Copying Genius reproduction free of disease.
- Production that produces roots potato tubers, while spared the same plant tomatoes.

- Production of human beings does not depend on food animals, but on solar energy and carbon dioxide since it contains chlorophyll.

In the area of electronics and computers industry witnessed the following:
- Computers that help the doctors in the future to conduct the necessary tests, such as analysis of blood and others, then diagnosis of disease and provide medicine for patients.
- The management of entire houses by computers, especially those inhabited by people with disabilities.

In transportation and communications, the human beings will benefit in the future from doing the following:
- Doing his own research without going to the university, but even before his computer.
- Attending the scientific conferences held thousands of kilometers away and participates in the discussions without the need to be present physically.
- Weather forecasting, prediction riches buried within the earth.

In the area of space, scientists also forecast to do great achievements at the level of outer space, including:
- The establishment of settlements in space, the moon and Mars, and the atmosphere surrounding land.
- Establishing satellite factories producing many electronic components and medicines.
- Conquest of outer space and freezing of embryos then placed them on the spacecraft, which will pass through space.

In the future, the peoples will be able to participate in the materials industry to produce materials with distinct qualities that can be used in the industry of cars, garment, as well as the use of carbon instead of silicon in the manufacture of bio-computers, and in other areas, there will be success in the war against the diseases, scientists expect that they will know the secret of the cancer, then to destroy it, also to understand the old age, and then to find the means to prolong life.

In about two hundred years, there will be four possible scenarios for future reference, namely they are:

- Very pessimistic.
- Cautiously pessimistic.
- Cautiously optimistic.
- Eager for growth and technology.

In future new generations will be created and involved, the so-called future shock and Toffler goes through analysis that includes:

- The future shock is a severe illness that affected increasing numbers of human beings and can be called satisfactory inability to adopt to rapid change.
- Reactions to the future depends on what is known about the ability to adapt recalled hypothesis in the 1st. instance, that the proliferation of psychiatric and neurological a lot of people due to the shock of the future and technological progress and scientific facts have not been able to create awareness on assimilations.

The search on sciences started with the beginning of man himself. The primitive scientific knowledge was transmitted from one generation to another through various professions. During the late nineteenth century the scientific education was dominated by religious, descriptive character and centered in the church. The science education in the early twentieth century was characterized by an increase turnout with those of social and economic modest level with scientific study. But, during the mid-twentieth century, emphasis has been used to enrich the life of scientific research, according to the needs of society and the role of science as a social power. The curricula of science have been evolved and the reform movement has emerged in the seventies because of the need for curriculum development for scientific creativity. In the year (1983) a report was issued entitled the nation at risk in United States of America as a warning to the community. The learning of science should deal with many available factors such as principles and conditions, including early education of children in primary schools, middle schools, and develop the ability to use scientific method to solve problems and the acquisition of the scientific education. The science education should have a role in the vital revolution, the acquisition of facts, the acquisition of functional concepts of the scientific principles and skills to solve problems and the development of scientific trends.

The development capacity and the development of personnel skills have witnessed successful experiences in building human being in the world.

Human beings play a distinct role in the educational process and the educational system due first to the human activity and second to the component of humanitarian and social structure. Education in general inherent rights throughout the human being's life - the life of humans or living organism possess, what is referred to by: thinking, feeling, and the ability to speak, despite the existence of opinions and philosophies on the relationship between body, mind and soul. All the activities related to human life building represent a coherent body with a specific mental condition and functional.

Despite these perceptions, the human is building various scenarios in the philosophies of public and educational philosophy. The human being in ideal philosophy is exercises his ability and responsibility for his actions, whereas Plato stated the man consists of two parts, one belongs

to the world of ideals (self) and the other to the world of sense (the body) 6and stressed on the idea of bilateral. While realists see that man is an organic entity with social tendencies. The natural philosophy suggests that the human soul is good, the present is the origin of the evolution of the future and education is necessary as long as they continue to longevity.

Al-Ghazali as a leader of the Islamic philosophy indicates the interdependence of the dimensions of human nature (body, mind, soul, heart), as well compatibility and balance

The development capacity and the development of personnel skills have witnessed successful experiences in building human being in the world.

Science and life

There are several theories to explain the emergence of life in trying to answer the classic question. How did life begin? The answer is as such: the fall of some organic molecules on Earth from comets, and planting the land with life by the intelligent beings on advanced planets. The emergence of life after eons of consecutive nomination (years) with quick chemical composition following the ground composition .Then after a short time period (the escalation of water droplets to the surface followed by collected chemicals, turned quickly to life in other place, then moved to the ground). After the general review of ideas about the evolution of life, the question was raised about the secret of inductive creation of the universe and the concordance between the creation of the universe and life.

The search on sciences started with the beginning of man himself. The primitive scientific knowledge was transmitted from one generation to another through various professions. During the late nineteenth century the scientific education was dominated by religious, descriptive character and centered in the church. The science education in the early twentieth century was characterized by an increase turnout with those of social and economic modest level with scientific study. But, during the mid-twentieth century, emphasis has been used to enrich the life of scientific research, according to the needs of society and the role of science as a social power. The curricula of science have been evolved and the reform movement has emerged in the seventies because of the need for curriculum development for scientific creativity. In the year (1983) a report was issued entitled the nation at risk in United States of America as a warning to the community. The learning of science should deal with many available factors such as principles and conditions, including early education of children in primary schools, middle schools, and develop the ability to use scientific method to solve problems and the acquisition of the scientific education. The science education should have a role in the vital revolution, the acquisition of facts, the acquisition of functional concepts of the scientific principles and skills to solve problems and the development of scientific trends.

There are clear trends in the study of the secret of creation of the universe and life: firstly it is believed that there is an aim in it and secondly there is a deny of the existence of the end, but to the existence of chance. Among the topics that are interesting mystery of the creation of the universe and of life, origin of the universe, evolution and various theories that talk about evolution, which accept both science and reason, scientists and philosophers. As well as the emergence of life and theories of ancient and modern, material and ideal, west and east that deal with them. The evolution has contributed to clarify the continuing march of life regardless of the theories put forward, including Lamarck and Darwin.

Intellectual life of human beings

The scientific progress made by man is in fact the result of intellectual growth in the presence and the disclosure of the laws governing the universe after the dovetails of the theoretical thought with that at experimental. In the modern era we find that the traditional division between the old mental philosophy, which assumes that the mind is the source of knowledge and the experimental philosophy which considered the experiment is the source of knowledge is still not significant fading. However, we believe that the following ideas may help in determining the intellectual life of human beings:

- The reason is a large scale in human thinking.
- The experiment is an important tool for the application of standard mental, but not alone.
- The trend is assumed that the mental does not ignore the role of experience in science and human knowledge.

Model of the concept of a philosophical mystery of life

There are three philosophical concepts of the world, developed as a result of human intellectual effort, the spiritual concept of the real material and the realistic concept of the divine. These contents could be evaluated and trying to rush to one of them or the formulation of the concept of compromise between them. The conflict between the divine and the physical manifestation of the conflict between idealism and realism and the philosophical conception of the world by one of two things the ideal and the concept of material does not correspond to reality at all. Realism is not in accordance with the concept that the ideal material as it is not the only thing that objects the concept of material, but there is another concept of realism that is the realistic is the concept of the divine. The concept of the divine world that does not mean dispensing with the natural causes, or something to rebel against the facts of science but is the correct concept of God, which is a deeper in reason. The material back in the door is the ideal claim to the area of spiritual, either in spiritual theology is a way to look at the reality.

In the scientific field, there is no divine, and the philosopher, whether divine or material believes in the positive side of science. There is no question of divine, according to philosopher of scientific material, but there are two incompatible and when it was the question of existence beyond nature.

The Divine believe that the world is the fact that just about to rule, outside the experiment and material, deny it believe to be natural causes that revealed by the experiment and spread to the hands of science that is the primary reasons for existence.

The nature of the evidence that can be given by Divine that is mind, not by the direct experiment, unlike the material traditionally regarded in the experiment as defined evidence of the materialism which is represents their own version. Believing in the concept of divine or metaphysical issues in general can not be proved experimentally. Materialism need to provide proof on the negative side, which distinguishes itself from divine and it is philosophical trend similar to divine, because science alone does not prove the concept of the physical world to be a scientific material, but all what is revealed by the science are the facts and mysteries in the natural world that leaves no space for the presumption of reason over the top of the material. If we look at a set of basic concepts of life and way of thinking, we could be addressed first to the theory of knowledge and secondly to the philosophical conception of life. The focus on mental dependence of the way of thinking includes knowledge over experiment, as well as a valuable study of human knowledge on the basis of sense mental sense, not physical.

Theory of knowledge and secret of life

The vision and ratification, which represent the expressing of the perception .The first represent a presence, such as heat, light and sound while the second (ratification), which represents a recognition that, for example heat is an energy comes from the sun and other concepts has dealt with a number of theories of perception, including:

- The theory of recollection

The (recall the previous information and the human soul exist separately from the body), it is isolation from the material, and we can correct some mistakes of this theory which is that the self is not something that exists but an abstract before the existence of the body, but of the fundamental movement of the material.

- Mental theory

This view put forward by Decarte, Cant, pointed to the existence of two perceptions of first is the sense of (heat, light, taste, etc...) and the second instinct (the human mind has the meanings and perceptions did not emerge from the common, but are fixed at the center of instinct). Differences with the theory of recollection is representing the fact that the senses the source of the perceptions but not the only reason. A disadvantage of this theory is attributed entirely to the sense of perception.

- Theory of sensory

This theory is based on the experiment and the sense which is the infrastructure of which this theory and the base of the human imagination. Therefore, the theory of knowledge could be used to follow-up to the secret of life and access to it .In this perception (scenario of objected value) of life is an expressing the presence of thing in our brain. Certification is which reveals the objective existence of biological perception, the mind feel the need to believe in it.

- Scientist of physics, biology collect together and living together at first, and other manifestations of heat and other, then start applying the principle of mind that it is necessary (the principle of the attic) which is the view (that the reason for every event) and then the mental principles is the general basis for all scientific facts and the principle of relativity.

When talking about the mind to explore the mystery of life it is necessary to underscore that the measurement is the first of the human thinking, which is in general for being processed cards dealing with enriching human thought with energy beyond the material. The intellectual track is gradually moved from public issues into private and the rationalism does not ignore the role of experiment. In the experimental doctrine which depends on how empirical features emanate from the part to the whole and the thought of public to private. Newton, for example, introduces a general law of gravity in the light of experiment, he has not felt that power through his five senses, but explored it by another phenomenon. He did not find explanation, but through the presence of the force of gravity, it was felt that the cars are not going in a straight line, but in a spin turnover.

The science and thought in the understanding of life

After clarification the thinking part of life, we could say the deeper main reason of the universe and the world in general is the reason due in particular, which ends by the sequence of causes and the only question that deserves its generation is due to this reason in particular, which is the first fountain of existence, is it the same material, or something else over the border. The philosophical trend of the mentioned concepts raises a question of whether the reason is the driving force of the world, is the same material and whether the secret of life is something else, beyond the limits of material and different from them. The philosophical conflict defines the direction of the secret of life; it requires not to confuse between the scientific material and the philosophical material.

Science and heritage

Each community the public and the private has a scientific legacy, with various pillars such as cultural, social and preserving the heritage is a national and human necessity. The scientific education can contribute to the benefit of the scientific heritage. So it is really important to stop a bit to show how to deal systematically with this legacy, thereby the process of transforming extrapolation of tradition and its benefits to the renewable power of the development of general education. There are methods and appropriate methodologies used for the purpose of benefiting from inherited science, including:

- Scientific heritage is treated with inherited methodology based away from the indiscrimination and excessive, damaging to these inherited disciplines.

- Scientific heritage depends on inherited as a single bloc in terms of time and the researcher does not fall blackouts, at the wrong focus on specific topics.

- The researcher shall comply with the spirit of scientific criticism of the inherited and not influenced by intolerance and prejudice.

- Selection process in the inherited case depends on modern scientific techniques.

- The researchers, scientists and leaders do not deny the human rights of those discoveries which have enriched as:
  - The invention of writing by the Egyptians and Sumerians.
  - The invention of printing in 1456 by the Germans.
  - The invention of the computer by the British and the Americans.
  - The invention of the information revolution by Bill Gates.

Models of inherited science

Scientific heritage is the most precious legacy of the human to human being and the complete presence of human that provide him with the meanings that make sense for a human to distinguish them from other organisms, however at the Arab and Islamic library of hundreds of men did not lift them up after the dust of oblivion, with great scientists of every color and every jurisdiction, those involved to raise the Arab Islamic civilization. These scientists are all waiting for their turn to come to study the scientific and objective research that must be available for each investigator and researcher.

People are beginning to rethink of their past, began to see in this last some of what sets for the process of starting again. In spite of what has become rich in the Library from the Arab-Islamic values to make it match the libraries property of the other principles. Many of the Arab and Muslim scientific heritage is still the reason for submission modern style and the modern curriculum.

The period after the Islam, which has widened the prospects of Arab scientific movement and the consequences of this period emphasizes inherited a multi-faceted so-called Arab Islamic civilization that include the following:

- Profiles of the scientific legacy.
- Islam and its impact in the scientific tradition.
- Holy Quran as a model
- Scientific thought in Islam.
- The translation of science.
- Methods and ways of scientific research in the scientific heritage.

Knowledge and science

Knowledge is the product of mental process of education and thinking. It is the basis of force and earnings constitute the most important components including the work and activity, particularly in the relation to economy, culture and education, which is also an economic asset. Furthermore, it includes the human being and his care and preparation of the main sources of knowledge.

Others believe that knowledge is the awareness and understanding of facts or acquire information through experience and means to acquire the unknown and self-development and is directly related to information, education and communication. The knowledge is also the most important component included in any work or activity, especially with regard to economy, society and culture.

The World Bank report entitled "knowledge is the path of development" explains the knowledge and conditions, the gap between North and South at the end of the twentieth century shows that scientists in the world are unevenly distributed between North and south and as follow:

- Africa   0.7%
- Arab States   1.5%
- North America   19.8%
- Europe   20.2%
- Asia   32.4%

Moreover, the knowledge society is characterized as one of the most important product or raw material. The knowledge of modern technology is used in their community and not to be complaint about to the presence in the same geographical location. It became one of the most important components of capital in the current era which is supposed to be free; free application for the benefit of community and should remain free.

The transmission of knowledge of the individual is called learning; the process of receiving knowledge through study is called education, which is the process of making the learner acquire knowledge and skills. The university exercises education and the student is learning.

Knowledge Society

Knowledge society is defined as a range of convergent interests, trying to take advantage of pooling their knowledge in areas that are interested. The knowledge society is of post-modernism linking the knowledge economy, generates profitable commodity.

Knowledge society requires the potential and special skills and abilities, super sophisticated infrastructure, natural resources and minds capable of producing knowledge and converting it into super economic progress.

The challenges of knowledge society emphasize the importance of university education, and this requires improvements in university education systems in order to transform them into the production process of knowledge.

Higher education is the foundation factor in the relationship with the knowledge society, which facilitates all exchange processes. The advanced stage of acquiring knowledge is the higher education which is the wider entrance to the knowledge society.

Knowledge society, which is called the community of the twenty-first century (third millennium), has the potency for production of knowledge and converting it into profitable commodities which led to the strength of this community and adopt a set of values including:

- Intellectual flexibility.
- Team work.
- Adoption of democratic values.
- Promotion of intellectual diversity.

•

The infrastructure of "knowledge society" includes:-

•   Physical infrastructure.
•   Technological infrastructure.

These infrastructures denote the facilities, services and logistics needed by the community such as, means of transportation and means of communications.

It is worth mentioning that the knowledge society is the basis of the information society, according to the report that was issued by the United Nations educational scientific and cultural organization "UNESCO" in 2005 under the title "From the Information Society to a Society of Knowledge".

In Iraq, the knowledge society is defying Iraqi universities for being easy to be challenged, since knowledge society is advanced, and Iraqi universities are backward. Entering the university into knowledge society requires providing sophisticated infrastructure of communication technology, favorable climates of stability and contemporary education system with new techniques that emphasize the supreme actual operations.

Production and dissemination of scientific knowledge

The production of knowledge is an advanced stage that acquires knowledge and can be measured in this production through scientific publications, patents and innovations. As noted, a significant increase in the movement of scientific publication in the Arab countries, where the number of publications of the Arab scientists elevated annually from 465 bulletin per year during 1976 to about 7000 paper in the year 1995 in different scientific fields such as medicine, agriculture, basic sciences, according to the human developments report for the year 2003.

It was noted that a wide gap between Iraq and the nations of the world in the production. In order to reduce this gap substantial changes has to be carried out in the education system especially the universities, where there is a need, to build up, a modern communications network, the rules of knowledge, new information, participation in the information systems and disseminating a culture of team work education, the fingertips of the Ministry of Higher Education strides for the insurance of scientific journals in various specialized scientific disciplines and humanity.

The patents registered in the United States of America for some counties during the period 1976-2002 differ, where the numbers of such patents for South Korea 27298 and Sweden 26318, while others have 104.

The number of workers in the production of knowledge is rising steadily but there are weak response to the demands of the market in higher education systems of Arab countries and shortcomings in the education requirements in the fields of science and culture.

Scientific Publications

There was a scientific awakening without any producing of knowledge such as dissemination of science in which the latest scientific achievements in the whole world can be found.

The discussion on scientific research publishing is extremely important, because at this stage of the development of science it plays very effectively to the enrichment of scientific research, expanding its scope and providing them with the achievements of researches, as well as to compensate for the lack of specific activities and creative achievements.

Methodology of Scientific Publication

Methods of scientific publication include the following:-

- Specialized scientific publishing.
- Public scientific publishing (radio, television, press, popular magazine)
- Scientific documentation (scientific publishing with confident nature).

The specialized scientific publishing gives special character followed by systematic nature and scientific method in writing, and is exposed in details to the process of scientific research and technical complexities involved and many difficulties and problems. The important point is supposed to disclose an interest in the reality of specialized scientific publishing as an essential step to improve this situation and its development.

The diagnosis of the problems of scientific research is the basis for the introduction and description of their treatment, ensuring the interest of achieving scientific research, which constitutes an integral part of publishing scientific research.

In the light of existing problems, for example, there is above all an urgent need to develop standards of sophistication and effective legislation to regulate the process of scientific publishing and editing of improvisation according to:-

- Documenting the work that had been published, at the level of each country
- The need for providing moral and material incentives.
- The role of the global information network (Internet) as a significant means of access to scientific information.

There is no doubt that the improvement of publishing scientific research should not be limited to specialists but must deal with the public scientific publishing (press, radio, television and scientific journals).

Global Scientific Publishing

Scientific publishing is concentrated in industrially and economically advanced countries. Nine countries produce 90% of the global scientific production registered by scientific information institutes. The scientific production of the developing countries is 6% of the international scientific production, such as:-

Asia continent	3.74%
Latin America	1.15%
Middle East and North Africa	0.59%
Black Africa	0.38%

It is worth mentioning that India then China are pioneer in production of scientific publication and have maintained a rate of 1:5. Egypt has advanced position in the list, despite the apparent increase in the quantity of scientific production in developing countries.

The comparison of the rates of publications of domestic and external level, in different continents, will find out that researchers of East Asia and Latin America are publishing more locally, whereas in Africa the publications high abroad. In developed countries the publication abroad does not exceed 20% for France, 25% for Japanese, and 12% of the overall European researchers.

All periodicals published in languages other than English are not worthy deserving the attention. The statistical reports that included researches from English speaking countries show that very low percentage published in other languages, while it is estimated that 17% of researchers published in English for French-speaking and 36% of Spanish or Portuguese-speaking.

Scientific publishing in developing countries

Studies that rely on international banks data incline to divide researches of the third world into two categories:

- Those who promote their products abroad in international periodicals with specified weight are taken in consideration.
- The second category is those who submit a local science without great value. The weakness of this provision despite the wide spread and the local science is not necessarily low-quality.

It was evident from other studies, conducted at the East Asia countries, where local scientific journals had reached a level of sophistication and researchers publish with options and not because of inability of external publishing.

The articles of researchers in developing countries received much attention when published with researchers from industrialized nations. This is a very important point that is the choice that must be carried out by researchers from the third world research in science that prevail globally and the trend towards solving local problems, even if it was not expected that this research sheds worldwide attention.

All periodicals issued in the third world do not receive the attention, whereas
- Among 201 chemical periodicals, Brazilian institute of information rely only on six.

Scientific Research

Scientific research base as the foundation for development and is divided into developed and developing countries, including investment in scientific research therefore, requires the development of plans for internet in scientific research and development for the purpose of narrowing the gap with advanced countries by allocating appropriate annual budget of up to 1% of national income available through the effective administration and legislation.

Scientific research has contributed to changing economic and social features of the peoples of the world and the contemporary world to total states vary according the level of economic, educational and scientific level. They are in four groups, as follows:

Group 1

Countries that create science and technology that are located in the highest degree of peace and science, including:

- USA
- Japan
- Western European.

Group 2

Countries introducing elements of scientific progress are:

- Russia
- Some countries in Eastern Europe

Group 3

Countries introducing elements of scientific progress are:

- Some from Arab countries
- Southeast Asia

Group 4

The countries that do not possess any means of scientific progress, the consumer and importer of scientific research outputs include:

- Most Arab states
- The rest of the world

Scientific Research in Iraq

Interest in scientific research began to change after the concept of university education, so that teaching is no longer the sole function of universities. However, the insight into scientific research in Iraqi universities shows that scientific research suffers from many problems including lack of funding, as the salaries of members of the teaching staff amount to nearly 90% of the budgets of universities. The remaining is distributed on all other aspects of spending, and thus negatively, it affects the requirements for scientific research devices and other supplies, as w as other sufferings of scientific research in Iraqi universities. It is considered merely functional performance by teaching, to achieve the goals of self separate from the research.

It is clear, that reality of scientific research in Iraq does not live up to expectation of a country trying to do in turn to contribute to the progress of scientific research, access to higher share of the global economy, investment of knowledge as a result of scientific research and development. There are systematic   obstacles   and   constraints   linked   to legislation and administrative, other impediments related to financing of scientific research and those relating to human cadres and constraints related to the interface and communication.

Methodological constraints is related to education in the methods of scientific research, while those associated with legislation is a clear view of the fact that instruction of promotions, for example, encourage individualism while those of the administrations found in the department, colleges and universities, the ministry and specialized research centers suffer from lack of coordination between different departments of research. The Obstacle of funding is one of the main channels in the country which is much lower than the proportion of GNP, representing one of the most major constraints for scientific research. While the cadres constraints refer to the wishes of the students involved to obtain certificates to improve their livelihood rather than being concerned with scientific research and increasing burden of teaching to teach in universities and the lack of clear criteria for selecting cadres which assume responsibility for scientific research centers.

A sound scientific planning for the development of scientific research of the university requires first the study of the current reality as it is a scientific study designed to learn the possibilities available in the universities and was done in this area to identify the negative aspects of it.

The scientific research at universities in Iraq, is one of three core functions of teachers as well as teaching and community service is linked with the promotion of academic holders of doctorates, while the remaining percentage of teachers holding a master degree of the operation full-time faculty.

There is a constant increase in the number of teachers in Iraqi universities that commensurate with the continued increase in the numbers of students enrolled in universities, and the indicator on the high ratio of students to teacher, an additional burden on teaching, which in turn reflected negatively on scientific research quantity and quality in universities.

The postgraduate students in Iraqi universities are an important part of the manpower involved in scientific research. Those students shape the future researchers, at universities in developed countries and have an important role in the completion of scientific research. Despite the expansion of universities in postgraduate programmes in Iraq, their contribution to scientific research is still pursuing their studies and specification in difficult circumstances. The current reality of manpower in the relevant scientific research in Iraq is characterized by low numbers of full-time researchers in spite of individualism to conduct research and the scarcity of complementary research teams.

The high proportion of the number of students compared to the teachers of acknowledged international figures, and the scarcity of opportunities for research assistants and technicians for training in developed countries to deal with the specialized instruments and maintenance of research laboratory, is a problem that should be considered seriously. The available statistics indicate that Iraq had spent a few sums of money on research and development after (2003). In an analysis of the budgets of universities refer to focus on supporting research projects, publishing and missions.

It is well known that the funding of scientific research in a university in developed countries comes mostly from the industrial sector and has reached the proportion of the funding year (1996) in countries like Japan to 67% and the United States to 63%, while the contribution of the industrial sector in supporting scientific research in Arab countries is very modest, almost Non -

existent in many universities. Statistics indicate that the total expenditure on research and development at universities and colleges in America (2002) to 36,333 billion dollars when the federal government contributed.

The productivity of doctorate-holding teachers in Iraqi universities in (2003) was 0.44 researches for each holder of a doctorate each year, while it was 1.00 a year ago (2003). Ministry of Higher Education in (2002) in collaboration with universities strides for the issuance of specialized scientific journals in nature and in various scientific disciplines and it is hoped that these magazines with a scientific level are classified and adopted globally.

The nature of research and investment that may lead to creative results after the year (2003) and industrial applications and patents are rare. It is clear, that contribution in this area is very modest, because of the inability of the institutions of higher education to transform the results of scientific research to investment project because of several obstacles. Iraqi universities after the year (2003) have received a package of legislations that have given special attention to scientific research. Instruction were issued own encouragement, regulations adopted by all universities and research achievements as a condition for promotion of distinguished professor status. The legislations included those issued by the university foundations to support scientific research, publishing support and instruction to published journals by the addition to patents and instruction: for the dispatch of teaching attending conferences and symposia.

Chapter Two

Pioneer sciences

- **Preface**

Splitting of the terms of reference for many Pure Science in chemistry for example, given the competence of Chemistry in the industry next to the jurisdiction of the Medical Chemistry

These terms of reference have been developed in US universities in physics, chemistry and mathematics to prepare graduates in some sectors, such as chemistry, physics, engineering and medical chemistry.

- Some of the topics were added assistance to many of the terms of reference of Pure Science and Education, including education and literature and Library Services and the use of modern equipment and robots and computers.
- Specialties in British universities through the expansion of preliminary to benefit from an increase in college of teaching systematic and practical opportunities and qualify students to work in various sectors of production.

To provide students with the qualifications can be expressed as follows:
- A sufficient culture to qualify them to understand the difficulties.
- Encouragement self-confidence of all to address the difficult technical issues.
- The amount of uncertainty makes them ask the right questions.
- The amount of persistence makes them continue to search for appropriate responses.

The scientific progress made by man is in fact the result of intellectual growth in the presence and the disclosure of the laws governing the universe.

In modern times, we find that the old traditional division between mental philosophy that assumes that the mind is the source of knowledge and the experimental philosophy, which is the source of knowledge

- Science can be divided into different sections, including the basic (Pure) and applied humanities and basic sciences include the natural sciences with the exception of Engineering and Applied Science, that include mathematics, physics, chemistry, life sciences, earth sciences, astronomy and meteorology

-

-

Some US universities has expanded dramatically in granting bachelor's degree, in the disciplines of chemistry related to medicine, agriculture, engineering , life sciences, physics, education, and others. Mathematics:

It is worth mentioning that the evolution of mathematics inherent in physics, then ripped off mathematics and liberation and began the development Systems abstract using set theory and mathematical logic, and this, using the workers and researchers in mathematics, and ways to create a broader understanding of the economy and population change., adoption of the so-called mathematical modeling introduced then "applied.

In the US and British universities from the beginning of the twentieth century and to the previous years at the end of the twentieth century sections of Applied Mathematics and other pure mathematics has become at the present time for Mathematical Sciences.

Mathematics speeds varying and expanding its uses and trends in the areas of knowledge, including the construction of mathematical models, and collaboration with life scientists and members of the organism and to understand human behavior and medicine to build mathematical models to help man in control of some natural phenomena, and installation constructivist   of   DNA. (Catastroph Theory) which can be used to describe some phenomena is continuing and is a multiple features, including applied and philosophical side, it can be adopted in the life and physical sciences

The world witnessed in the twentieth century breakthrough in all fields and scientific trends, so there are no boundaries between different disciplines.

For example, medical science requires engineering science and recent tests of modern science depends on the physical, chemical extraction and analysis and also relies on mathematics to lay the groundwork mathematics.

The progress and development in pure science, for example, develop new subjects and disciplines and specialties of science. New interfaces were not known during the first half of the last century. The results of these major changes in curriculum and build up research transformed these developments to the university curricula. Seminars and researches are now carried out in different ways including:

- Bachelor based on the study and theses.
- Some universities in Britain and Germany developed curricula at the level of initial studies that include research and study.
- Divide the present fields such as industrial chemistry, chemistry of life with medical side and other disciplines in the branches of pure science.
- Develop competencies.

These terms of reference have been developed in American universities in physics, chemistry and mathematics, to prepare graduate in some sectors such as engineering, chemistry physics and chemistry, agricultural engineering, and medical studies.

- Adding assistance topics.

Some topics have been added as assistance of many of the terms of reference of pure sciences, including education, literature and library services and use modern machinery and computers.

- Other disciplines (Sandwich)

British universities were carrying out by expanding the initial years of university study for use in increasing opportunities for the systematic teaching and applied for and rehabilitation work in various production sector. Recent scientific trends in Iraq.

assistance topics.

Some topics have been added as assistance of many of the terms of reference of pure sciences, including education, literature and library services and use modern machinery and computers.

Other disciplines (Sandwich)

The recent trends in the chemistry of life ,Bio- chemistry studies the chemical and physical characteristics of the components of the cell and features of the life systems of the components, as well as the interpretation of what these systems in the cell Biochemistry provided a lot of accomplishments, it has helped to clarify the mechanism of medicine and contributed to the diagnosis and treatment of many diseases.

-

- Each country has special methods to determine its own disciplinary university and identification numbers, graduate students and the quality.

- The world witnessed in the twentieth century breakthrough in all fields and scientific trends, so there are no boundaries between different disciplines. For example, medical science requires engineering science and recent tests of modern science depends on the physical, chemical extraction and analysis and also relies on mathematics to lay the groundwork mathematics.

- The progress and development in pure science, for example, develop new subjects and disciplines and specialties of science. New interfaces were not known during the first half of the last century. The results of these major changes in curriculum and build up research transformed these developments to the university curricula.

- Other disciplines (Sandwich)

- British universities were carrying out by expanding the initial years of university study for use in increasing opportunities for the systematic teaching and applied for and rehabilitation work in various production sector.

- **Chemistry**

The splitting of the many Pure Science such as chemistry formed a new terms such as industrial, medical chemistry.

The introduction of these terms in the international universities in physics, chemistry and mathematics to prepare graduates for some sectors such as engineering and physics engineering, medical and agricultural .

This was done in British universities through the expansion of preliminary to benefit from an increase in systematic college teaching and practical opportunities and qualify students to work in various sectors of production.

Some universities introduced in the world of modern curricula at the level of initial studies, as in Britain and Germany, for example, in the field of chemistry especially during the second half of the twentieth century, dealing with developments chemical industries.

Biochemistry lasted over the age of a century in different disciplines, some with a study of the materials that make up plant cell and then called the chemistry of plant life, and then which is related the animal cell which is called chemistry of animal life if the human cell is the target.

Chemistry has expanded to clinical biochemistry that includes chemistry of life, becoming a physical, organic and biochemistry and inorganic chemistry as well as nutrition. Interested in chemistry, life functions of the modern systems of life, have contributed to the means of study in the last century with the observation of these systems directly during the work,. Either at the present

time which has changed the picture and it became possible to obtain the most desirable observations by the development of viable technologies (electron microscope, radioactive isotopes.

The scientists believed at the end of the nineteenth century that it is possible to obtain some information relating to the systems of life, by studying the chemistry of cells and for decades was followed by chemists adopted the chemical methods available and succeeded in obtaining useful developments.

Significant improvements to the technical methods such as the use of chemical isotopes have greatly increased the sensitivity of diagnosis of different types of molecules of life and others, and when it is necessary to separate the components of the chemical reaction through life and is very sensitive, then used deportation electric traditionally.

When the attention has turned physicists, chemists, physicists about the science of life (and perhaps due to the ability of living cells to configure the system, although the laws of physics, emphasizes the universe there is a tendency towards non- attendance) then emerged the technical methods of physical, chemical, physical,.

The progress achieved in the chemistry of life has begun to acknowledge that the livelihood systems containing small particles interested m organic chemistry to study and clarify as well as large molecules called macroscopic particles which are not molecular weights less than 100 million times the mass of one atom of hydrogen..

The importance of macroscopic particles of the life systems in its ability to privacy in life interactions composition of building blocks, and can say clearly that he had made in the past years considerable effort to characterize the annexation of macroscopic particles as well as the reactors that occur between them and the need for advanced methods of separation and purification and characterization of macroscopic particles.

.

The objective of biochemistry for nearly half a century is to collect and organize interactions that occur in living cells. The motivation for this major effort is that a significant number of the attributes of living cells can be understood through these interactions that are typically characterized by the formation or breaking covalent bonds.

It is been clarified on the liberalization of energy as a result of break chemical transformation processes as well as molecules of life and mutual assembly operations amino acids, sugars and fats to form macroscopic particles.

During the last thirty years clearly demonstrated that the reactors that occur between molecules due to physical, those that are not or break covalent bonds have the same importance of chemical reactions, for example. The organization of chemical reactions (ie, the degree of permitted them to occur) performed by the physical changes that occur in the structure of large molecules, as well as the creation of active centers in these molecules and the resulting interdependence of the non-covalent small molecules, in addition to, many of the qualities of a macroscopic aggregates molecules in cells .

Plurality of molecules of life structure consists of installation of the first structural molecular structures of multiple different types of units place (serial), for example, the sequence of amino acids found in proteins and sequence by chemical analysis. The secondary structural composition involves the formation of a complex three- dimensional structures is called to direct all of the units for multi-particles to other units and is called the secondary structural composition tradition or (body and image) or the status of the foundation structure or backbone of multiple chains.

The forms, which consists of surfaces and different types of these mixed forms, and called on the direction of (position) of side chains relative (amino acids, nucleic acids or bases) triangular structural composition. A lot of multiple molecules of life with each other to be as complex as the structures of several structural units viruses, membranes and capillaries bonds and are usually in one level, where you specify the types of bilateral structures of proteins. On the other hand that includes the alpha carbon to allow for many types of structural combinations.

The two phosphate ester bonds in nucleic acids are subject to sag as well, because the flexible rule and hate water and one level surrounded by a few of so they are usually located one above the other, thus reducing the adhesion of water, and this increases the structural rigidity of installation.

The multi- life linear molecules, which has no free rotation about the bonds, which do not interact aggregates side is called the file is not a random combination structurally specific dimensions or size of distinct wraps by Brownian movement. Size can be measured by the value equal to the rate of rotation of the radius around a point or an axis.

Nucleic acids - the mystery of the mysteries of life

A nucleic acid represent the brain of the cell brain cell with a specific developed program, to be issued through the instructions for that cell fusion and installation of the life and death and plan for the future.

There are two types of nucleic acids (DNA, RNA) both of their differences centered a long chain molecules composed of nucleotides and position of certain forms.

In 1953 Crick and Watson was able, who have previously received the Nobel Prize in 1962 developed a model for the DNA structure, consisting of two strands of units of the four nucleotides arranged in orderly fashion, and every one of them is a multi-helical nucleotides wrapped around a common axis to form the double helix right direction.

The DNA double helix showing a right handed or B-helix. Indeed, this diagram is the only one that actually appears in the article, and one would seek in vain any diagrammatic elaboration of what the molecular structure of DNA is (although components such as deoxyribose or guanine are named as such in the text).

Anyone seeking to repeat Watson and Crick's model building would certainly have to acquire additional molecular data from another sources. Some of that missing information although this only describes the connectivity of the various atoms in a single strand of DNA, and not the two or three dimensional relationships of the individual atoms.

The molecular basis of one strand of DNA, based on the CG bases.

Armed only with the one diagram actually published, curiosity might lead one to pose a scientific question such as "How did Watson and Crick assign the helix as right rather than left handed".

In other words, on what data did they base that conclusion in most cases certainly erroneously Coincidentally, similar issues of left or right-handedness were to be found when Pauling presented his α-helix models of proteins. In fact, almost all protein helices exhibit right-handedness .

that both chains follow right handed helices ...

because left handed helices can only be constructed by violating permissible van der Waals contacts.

We are informed that such permissible contacts include the approach of any two hydrogen atoms in the molecule to a distance of no less than 2.1Å.

We are not however informed what the violations might be in a left handed helix that excludes this model. In other words, just how close can two hydrogen atoms separated (for intramolecular contacts) by at least four bonds approach? In fact, distances of ~1.85Å or less have been observed .

In this same full article by Watson and Crick . we are given a table of numerical (polar) coordinates describing the positions of twelve key atoms, but it would have taken a very determined scientist to have used only this combination of information to easily confirm the assertion that a left-handed helix is excluded.

Perhaps the lack of a model with which the reader could experiment might account for the relatively slow recognition of the importance of this article in the immediate years following its publication, and the observation

47

that whilst a physical model of DNA had of course been built, it was only available for viewing.

One tool that modern chemistry now has at its disposal (which Watson and Crick did not have) are accurate molecular models based on quantum mechanical calculations. Such a molecule is quite a challenge to model, since the computation has to take into account subtle interactions such as dispersion (long range correlation) effects, which are more or less equivalent to the van der Waals contacts referred to by Watson and Crick.

The ionic phosphate groups, the planar bases and how they stack, so-called anomeric effects at the base-sugar connecting C-N bond, hydrogen bonds between both the obvious NH...N and NH...O atoms and less obvious ones such as C-H...O, and not least the capacity to deal self-consistently and accurately with the optimal positions of (at least) 250-254 atoms.

In reality, such models have only very recently become available .

The models explained by x-ray, have two sections through the longitudinal axis of the first of 0,34 nm and the second 3,4 nm.

Model of "Watson & Crick Model" Both Watson and Kirk in 1953 the first specialist in genetics and the second physicist to develop a model structure that represents the structural basis of DNA. Jarkav in the light of studies on the percentage of nitrogen bases, X-ray dimensions, as well as the fact that adenine = thymine and Guanine = Cytosine

Features of the model of Watson and Crick, this is called beta-form and include:

The DNA is composed of two strands of deoxy nucleotides wrapped through multi-spiral system.

The nucleotide chains are connected by diester bonds within one strand.

The bases purines and pyrimidines facing each other, so that in particular, adenine faces thymine and guanine faces cytosine through hydrogen bonds.

The order of nitrogenous bases one in series vary from other.

The levels of sugar rings parallel to the axis of and phosphate groups abroad.

DNA is divide in two parts the first is called water hating (nitrogenous bases), located internally and a second which is externally faced of the surrounding water molecules containing phosphate groups.

The second half of the nineteenth century witnessed a series of discoveries of life such as, a serious (cell) theory at the hands by Matthias in the plant and then Theodor ishvan in the animal. Since of plant and animal are composed of cells, that evolution (cell theory) and their development is considering critical stage in the progress of life science similar to atomic energy.

The human Physiology science as one of the branches of the life science that refers to the amazing facts explaining the greatness of the Creator and accuracy of the details and secrets. The digestive system for example (the greatest chemical plant in the world) including, the methods of food analysis of chemical analysis of various surprising and distribution of fairly Safe food distributed to millions of living cells.

If we explore the science of life, we will find another secret of that biggest secrets, the secret of the mysterious life, which fills the moral conscience of mankind, with the concept of divine fear and faith, firmly established in it. The theory of self-regeneration was collapsed at the depth but the unequivocal scientific experiments, demonstrated the invalidity of the theory of self-regeneration.

The material basis of life science was examined and then basically spread the idea of elements. The atoms are spread better for the basic materials of the universe and second nature that the elements consist of a central core electrons of the nucleus orbit (negative) and the nucleus contains protons and neutrons. Attempts were made to alter the material to absolute energy, no electric charge.

In other words removing character from element in the light of the theory of relativity of Anstein, where the body mass is relative, not fixed, and increase with the speed according to Anstein equation energy = mass of × square of the speed of light and mass = energy ÷ square of the speed of light.

As a result, the atom, including of protons and electrons are condensed energy. Appeared in various forms and multiple images, whereas materials has been converted into energy and energy to the material.

It follows from the views put forward that the original materials the world-life the reality show one common in various forms, and the physical properties of compounds are accidental such as the liquidity of water is incidental, not self-evidence, since it is consisted of two toms and possible separation these two elements from each othe..

The characteristics of the simple elements themselves are not self-rule but are incidental to the material. That such material characteristics become the light of the above facts incidental, it is encroached to be among the identified energy and philosophically, the presumption of tahe material in the world of life on the top reason capable for denial, as well as of effectiveness.

.

The chemical side is concentrated in the molecules of life on the carbon, which constitutes about 50% by weight bio- molecules are characterized by life-covalent bonds, four of which related to carbon stubs and have different angles of particular value from one carbon atom to another in different molecules of life and because of that there are different types of building structures with three-dimensional, these structures contribute to clarify the complexity of the cellular composition with particular reference to its failure, as well as various forms. In addition organic compounds are characterized by free rotation. The tetrahedral bonds emphasizes on individual carbon atom of the very important properties of organic molecules and the presence of four different groups or different atoms connected to carbon and the last become non-symmetrical (a carbon-atom covalently bonded with four different groups) and composed (which every one of which is mirror images of each other) with a symmetric arrangement in space and called isomers mirror of light for the chemical similarity of the interactions but differ in physical properties of the rotation of polarized light.

Amazing developments occurred in the chemical sciences, especially during the second half of the last century, including implicit and other interfaces. Developments implied addressed the contents and mechanisms of new interpretations of the phenomena and events and chemical reactions, arose because of it also subjects new disciplines in within taught himself chemistry, led these developments as well as to open new channels in the search for science and technology, chemical and construction of new chemical industries, or chemical industries new technologies .

The developments of the second type of chemical interfaces science has dealt with disciplines that bind Chemical Sciences Pure and Applied Sciences and various led these developments to develop terms of reference or a new science were not known before and took these terms of reference, science its proper place in the programs and curricula and plans of study and research. Some of these updated science is still incomplete features at the present time, but the scientific forecasts indicate near the completeness or ingrained with the turn of the century.

And it reflected the effects of these developments both types of implicit and interface to the application of curriculum topics sandwich to work in various scientific and industrial fields. And it took some US universities to expand significantly in the granting of bachelor degrees, but also in graduate studies, in double chemistry disciplines dealt with medicine, agriculture, engineering or biological sciences, physics, education and others.

Attention to scientific research on the level of the initial study in universities to prepare messages also increased next to pass the courses. And the expansion and development of the two concepts in the event in the chemical sciences as well as to the valley bisected some specialties chemical known to specialists or more, and this would develop new terms of reference were not known before.

And it took international universities known enter the computer not only in the subjects of study specialized in chemistry, but also taking the means of clarification is necessary in the chemical t for the purpose of clarification and display structures and chemical structures and arranged stereo chemistry and methods of preparation and diagnosis of some vehicles, for the purposes of types of processors and accounts chemical necessary in the present day.

And the expansion of development in the chemical sciences led to the splitting of some terms of reference chemical known to specialists or more, and this would led to the development of new terms of reference were not known before. Amazing developments occurred in the chemical sciences, especially during the second half of this century, including implicit and other interfaces. Developments implied dealt with content and vocabulary and mechanisms known in chemistry and provide improved or new interpretations of the events and phenomena and chemical reactions. And arise because of it also new subjects and disciplines within taught himself chemistry. These developments also led to the opening of new channels of scientific research in chemical and technology to create a new chemical industries, chemical industries or new technologies.

The developments of the second type of chemical inter-disciplinary science has addressed that connects Chemical Sciences Pure and Applied sciences different. These developments have led to the development of the terms of reference or a new science were not known before. And it took these terms of reference and its proper place in science programs and curricula and plans of study and research in a lot of sober universities of the world. Some of these disciplines or updated science is still incomplete features at the present time, but the scientific forecasts indicate near completeness and ingrained with the beginning of the next century. And it reflected the effects of these developments both types of implicit and interface to the application of the topics curricula in some British universities to increase the rehabilitation of sections of chemistry graduates to work in various scientific and industrial fields. Some US universities and taken to expand significantly in the granting of bachelor degrees, but also in graduate studies, in double chemistry disciplines dealt with medicine, agriculture, engineering or biological sciences, physics, education and others. Attention to scientific research on the level of the initial study in universities has also increased even some British universities have become demanding preliminary study students in chemistry and other sciences to prepare scientific messages next to pass the courses. And the expansion and development of the two concepts in the event in the chemical sciences as well as to the valley bisected some specialties chemical known to specialists or more, and this would develop new terms of reference were not known before.

And it took universities in the world known computer enter not only study subjects specialized in chemistry, but also as a means necessary in chemical studies for the purpose of clarification and display compositions and structures and chemical and arrangements of steric and methods of preparation and diagnosis of some vehicles, for the purposes of conducting types of processors necessary chemical and accounts in the present day. Also have been vaccinated chemistry curriculum in universities in the world in a new Help topics it was not known before and believed it had become a day of themes and supplements the scientific culture

Some recent trends in chemistry and Bio-chemistry

Of the difficulties faced by the researcher to contain the recent trends in science variety and the limited field of view. Although it can review these trends very briefly and then focus on each other and thus the benefit is based effort to prevent COMPLICATIONS. One area of modern chemistry, for example, as follows: -

Macroscopic cyclic compounds.

These vehicles susceptibility to separate chemical elements and drawn with high efficiency, used as models to study the movement of ions across the cell wall and have a high efficiency as anti-life.
• multiple molecules (polymers).
therapeutic radiology It has a high ability to connect to the electrical elements and the movement of the medication to the exact location.
• Events interactions transition, to make an impact on increasing the effectiveness of life for the purpose of treatment compounds.
• Use life molecules of life for chip manufacturing computers in the next generation.
• Manufacture of organic chemicals super connectivity.
• The development of a new generation of super-computers and sensitive devices, can be grown within the human body to repair some damaged tissue.
• The development of spectroscopy devices to follow the pre-life interactions.
• Making new chemical molecules for the treatment of diseases (AIDS, cancer). Pave the recent trends in biochemistry
Looking biochemistry in chemical and physical properties of the components of the cell and the features of the systems life of these components, as well as the interpretation of the nature of these regimes in the cell in a minute.
Biochemistry made a lot of achievements, has helped to clarify the mechanism of drugs and contributed to the diagnosis and treatment of many diseases and provided techniques that could be used measuring a lot of compounds in vivo level.

Biochemistry century has different disciplines, some with the study of materials that make up plant cell and called then the chemistry of life and plant which relates to animal cell and if the human cell is intended whether natural or satisfactory called the clinical chemistry has expanded biochemistry, becoming a cover biochemistry Physical and organic chemistry of life and life Inorganic chemistry and nutrition as well as chemistry.

The progress achieved in the biochemistry has begun to recognize that life systems contain small molecules and clarified as well as the huge molecules called macroscopic particles which are no less molecular weights from 100 million times as much mass hydrogen atom. It Occupies important molecules of macroscopic Life Systems with ability towards life interactions in the composition of the structural units, and we can say clearly that in the past years huge efforts to characterize the annexation of macroscopic molecules as well as the reactors that occur between them and needs so advanced methods of separation, purification and characterization and to obtain information on structural composition of macroscopic molecule.

The aim of biochemistry for almost half a century is to collect and organize the chemical reactions that occur in living cells. The catalyst for this great effort is that a significant number of the qualities of life cells can be understood through these interactions that are usually characterized by the composition or breaking covalent bonds. It has been clarified as a result of the process of liberalization of the energy break-up as well as the chemical transformation of molecules of life and mutual assembly operations ofamino acids and sugars and fats molecules to form macroscopic processes.

Bio-security and the impact of CT on biochemical systems
There are specific levels of biosecurity have been developed in accordance with the rules set by the institutions (scientific, global) radiation enters, including in coordination with various other influences.

Form used radioactive materials in diagnostics, treatment and scientific research regardless of its source, whether a hospital or medical institution or research or the result of diagnostic and and nuclear medicine threat to public health and the environment, the same time by the right doctor in his battle against the disease.

The developments of the second type of chemical sciences interface had addressed the disciplines of science linking chemical sciences and applied various treatments. These developments have led to the developments of science or the new terms of reference were not known before.

The recent trends in the chemistry of life

Bio- chemistry studies the chemical and physical characteristics of the components of the cell and features of the life systems of the components, as well as the interpretation of what these systems in the cell Biochemistry provided a lot of accomplishments, it has helped to clarify the mechanism of medicine and contributed to the diagnosis and treatment of many diseases and provided techniques which could be used measure the level of many of the compounds in vivo.

Biochemistry lasted over the age of a century in different disciplines, some with a study of the materials that make up plant cell and then called the chemistry of plant life, and then which is related the animal cell which is called chemistry of animal life if the human cell is the target.

Chemistry has expanded to clinical biochemistry that includes chemistry of life, becoming a physical, organic and biochemistry and inorganic chemistry as well as nutrition. Interested in chemistry, life functions of the modern systems of life, have contributed to the means of study in the last century with the observation of these systems directly during the work, either at the present time which has changed the picture and it became possible to obtain the most desirable observations by the development of viable technologies (electron microscope, radioactive isotopes, Immunology, spectrum).

The scientists believed at the end of the nineteenth century that it is possible to obtain some information relating to the systems of life, by studying the chemistry of cells and for decades was followed by chemists adopted the chemical methods available and succeeded in obtaining useful developments. Significant improvements to the technical methods such as the use of chemical isotopes have greatly increased the sensitivity of diagnosis of different types of molecules of life and others, and when it is necessary to separate the components of the chemical reaction through life and is very sensitive, then used deportation electric traditionally.

When the attention has turned physicists, chemists, physicists about the science of life (and perhaps due to the ability of living cells to configure the system, although the laws of physics, emphasizes the universe there is a tendency towards non- attendance) then emerged the technical methods of physical, chemical, physical, such as spectroscopy, diffraction to be applied an the field of biology.

The progress achieved in the chemistry of life has begun to acknowledge that the livelihood systems containing small particles interested m organic chemistry to study and clarify as well as large molecules called macroscopic particles which are not molecular weights less than 100 million times the mass of one atom of hydrogen. The importance of macroscopic particles of the life systems in its ability to privacy in life interactions composition of building blocks, and can say clearly that he had made in the past years considerable effort to characterize the annexation of macroscopic particles as well as the reactors that occur between them and the need for advanced methods of separation and purification and characterization of macroscopic particles in order to obtain information on structural composition of the macroscopic molecule.

The objective of biochemistry for nearly half a century is to collect and organize interactions that occur in living cells. The motivation for this major effort is that a significant number of the attributes of living cells can be understood through these interactions that are typically characterized by the formation or breaking covalent bonds. It is been clarified on the liberalization of energy as a result of break chemical transformation processes as well as molecules of life and mutual assembly operations amino acids, sugars and fats to form macroscopic particles.

During the last thirty years clearly demonstrated that the reactors that occur between molecules due to physical, those that are not or break covalent bonds have the same importance of chemical reactions, for example, that the organization of chemical reactions (ie, the degree of permitted them to occur) performed by the physical changes that occur in the structure of (construction) of large molecules, as well as the creation of active centers in these molecules and the resulting interdependence of the non-covalent small molecules, in addition to, many of the qualities of a macroscopic aggregates molecules in cells or in the organism (the cell membrane and walls of cells and chromosomes).

Plurality of molecules of life structure consists of installation of the first structural molecular structures of multiple different types of units place (serial), for example, the sequence of amino acids found in proteins and sequence by chemical analysis. The secondary structural composition involves the formation of a complex three- dimensional structures is called to direct all of the units for multi-particles to other units and is called the secondary structural composition tradition or (body and image) or the status of the foundation structure or backbone of multiple chains. The forms, which consists of surfaces and different types of these mixed forms, and called on the direction of (position) of side chains relative (amino acids, nucleic acids or bases) triangular structural composition. A lot of multiple molecules of life with each other to be as complex as the structures of several structural units viruses, membranes and capillaries bonds and are usually in one level, where you specify the types of bilateral structures of proteins. On the other hand that includes the alpha carbon to allow for many types of structural combinations. The two phosphate ester bonds in nucleic acids are subject to sag as well, because the flexible rule and hate water and one level surrounded by a few of so they are usually located one above the other, thus reducing the adhesion of water, and this increases the structural rigidity of installation.

The multi- life linear molecules, which has no free rotation about the bonds, which do not interact aggregates side is called the file is not a random combination structurally specific dimensions or size of distinct wraps by Brownian movement. Size can be measured by the value equal to the rate of rotation of the radius around a point or an axis.

Nucleic acids - the mystery of the mysteries of life

A nucleic acid represent the brain of the cell brain cell with a specific developed program, to be issued through the instructions for that cell fusion and installation of the life and death and plan for the future.

There are two types of nucleic acids (DNA, RNA) both of their differences centered a long chain molecules composed of nucleotides and position of certain forms.

In 1953 Crick and Watson was able, who have previously received the Nobel Prize in 1962 developed a model for the DNA structure, consisting of two strands of units of the four nucleotides arranged in orderly fashion, and every one of them is a multi-helical nucleotides wrapped around a common axis to form the double helix right direction.

The models explained by x-ray, have two sections through the longitudinal axis of the first of 0,34 nm and the second 3,4 nm.

Model of "Watson & Crick Model" Both Watson and Kirk in 1953 the first specialist in genetics and the second physicist to develop a model structure that represents the structural basis of DNA. Jarkav in the light of studies on the percentage of nitrogen bases, X-ray dimensions, as well as the fact that adenine = thymine and Guanine = Cytosine

Developments in the synthesis and structural forms of DNA

Features of the model of Watson and Crick, this is called beta-form and include:

- The DNA is composed of two strands of deoxy nucleotides wrapped through multi-spiral system.
- The nucleotide chains are connected by diester bonds within one strand.
- The bases purines and pyrimidines facing each other, so that in particular, adenine faces thymine and guanine faces cytosine through hydrogen bonds.
- The order of nitrogenous bases one in series vary from other.
- The levels of sugar rings parallel to the axis of and phosphate groups abroad.

DNA is divide in two parts the first is called water hating (nitrogenous bases), located internally and a second which is externally faced of the surrounding water molecules containing

Amazing developments have taken place in the chemical sciences particularly during the second half of the century, including implicit and other interfaces. Developments on the implicit content and the vocabulary and mechanisms are known in chemistry and provide improved or new interpretation of events and phenomena and chemical reactions, as a result also of new subjects and disciplines within the science of chemistry itself. These developments have led to the opening of new channels in scientific research and technological innovations such as chemical industries to create new chemicals, or chemical industries, and new techniques.

The developments of the second type of chemical sciences interface had addressed the disciplines of science linking chemical sciences and applied various treatments. These developments have led to the developments of science or the new terms of reference were not known before.

The second half of the nineteenth century witnessed a series of discoveries of life such as, a serious (cell) theory at the hands by Matthias in the plant and then Theodorishvan in the animal. Since of plant and animal are composed of cells, that evolution (cell theory) and their development is considering critical stage in the progress of life science similar to atomic energy The human Physiology science as one of the branches of the life science that refers to the amazing facts explaining the greatness of the Creator and accuracy of the details and secrets. The digestive system for example (the greatest chemical plant in the world) including, the methods of food analysis of chemical analysis of various surprising and distribution of fairly Safe food distributed to millions of living cells. In view of these living cells the issue of causal efficacy and secret of life that fills justify the astonishment and admiration for self-

cell, while adapting to the requirements of their position and circumstances.

If we explore the science of life, we will find another secret of that biggest secrets, the secret of the mysterious life, which fills the moral conscience of mankind, with the concept of divine fear and faith, firmly established in it. The theory of self-regeneration was collapsed at the depth but the unequivocal scientific experiments, demonstrated the invalidity of the theory of self-regeneration. The material basis of life science was examined and then basically spread the idea of elements. The atoms are spread better for the basic materials of the universe and second nature that the elements consist of a central core electrons of the nucleus orbit (negative) and the nucleus contains protons and neutrons. Attempts were made to alter the material to absolute energy, no electric charge. In other words removing character from element in the light of the theory of relativity of Einstein, where the body mass is relative, not fixed, and increase with the speed according to Einstein equation energy = mass of × square of the speed of light and mass = energy ÷ square of the speed of light. As a result, the atom, including of protons and electrons are condensed energy. Appeared in various forms and multiple images, whereas materials has been converted into energy and energy to the material.

It follows from the views put forward that the original materials the world-life the reality show one common in various forms, and the physical properties of compounds are accidental such as the liquidity of water is incidental, not self-evidence, since it is consisted of two toms and possible separation these two elements from each other and the status of water disappear completely.

The characteristics of the simple elements themselves are not self-rule but are incidental to the material. That such material characteristics become the light of the above facts incidental, it is encroached to be among the identified energy and philosophically, the presumption of the material in the world of life on the top reason capable for denial, as well as of effectiveness.

Enzymes

The word of the enzyme was proposed by the researcher (kon) in 1878 and began to study the specificity of the enzymes at the end of 19th century. Fischer introduce in Fischer" in 1894 the idea of the specifity of the enzyme and the relationship between the enzyme and the substrate stereochemically.

The purification studies started on the isolation of enzymes and then purification in 1920 and in 1922 Wilstater has isolated some of enzymes Kdama and Dixon in 1926 extract an enzyme that oxidize Xanthin, and then the research was continued in developing the science of enzymes and its different branches. The chemical nature of enzyme was elucidated as protein, produced by the cells of the body according to the different needs of those cells. The enzyme as a catalyst differ from inorganic catalysts in the chemical nature, and mode of operation dynamics of interactions (mode of action), the need for special materials called Co-enzymes. The important properties to the living cell, its ability to catalyze complex reactions at temperatures of its environment without these reactions the cell activity occurs slowly, the cell room has thousands of enzymes and each enzyme specially designed to perform a specific reaction, according to a specific rule within the cell and the light can not imagine the existence of this huge number of specialized enzymes in an organism, which provides a way to control the cell chemistry as accurately as possible. "Oswold" has defined the catalyst as accelator of the speed of reactions, which is characterized by the following characteristics:

- It maintains the structure of the enzyme during the chemical reaction but may occur some natural changes in some of the reactions.
- Speeds up the reaction to reach the state of balance "Equilibrium" without prejudicing "Equilibrium Constant" or position, but also effect the speed of the reaction for each of the directions equally and keep the concentrations of various materials in the fixed chemical balance.
- Contribution to access to slowest energy

Activation Energy

This factor is characterized by the specificity of its reaction "Reaction Specificity", where there is usually an incentive, one for each reaction and group of close reactions.

Enzymes produced by various cells of the body according to the need for these cells. These stimuli differ from their counterparts the nonorganic catalysts by the chemical nature and natural way of working and the mechanism of the reaction that have driven. The enzymes which consists of protein are different from non-organic catalyst such as manganese dioxide, platinum, nickel, iron filings, etc, The enzymes are more specialized than non-organic catalysts involved in several reactions and be quite different from each other. The molecular weight of enzymes is large, more than the molecular weight of non-organic factors, the enzymes affected by temperature that increase their reactions.

"Esterases" catalyze the several types of esters in fat and phosphatides, varies its influence from ester to another. These enzymes do not affect one type of compound, but on many types which are similar in chemical composition in terms of the existence of the certain chemical bonds in all.
The enzyme "Lipase", which is one type of esterases specific in the analysis of fat and can not cataltze non-fat, esters. The relative specifity is not an enzyme specialized in the impact on a particular compound, but have specialized in certain chemicals, regardless of the building blocks are related to the chemical bonding.

Enzymes are specific for the space look-alikes "Stereo chemical Specificity" the enzyme is working on a particular compound and does not work in other particular compound, such as "L-amino and oxidase" affect the amino acid L but does not stimulate the conversion of the amino acid D.

"Structural Specificity" some of the enzymes required for the completion of their specialization, the existence of special groups adjacent to, e.g

"Carboxy Peptidase" affect on the bond adjacent to the peptide group &COOH free in many complex peptides. And the enzyme "Aminopeptidase" only affects the foreign peptide on the link adjacent to the free amino group of many peptides, affects the internal peptide bonds in the protein molecule.

Specificity of the reaction, the enzyme has the ability to choose one of multiple reactions (specificity of the reaction). The enzyme also stimulate one of the multiple reactions, through the catalysis of the specific substrates. For example, the enzyme "amino acid oxidase" oxidize and remove the amino acid "Oxidative Deamination" and removing CO_2 when there is an enzyme "Decarboxylase" and interaction that took place for the third amino acid is the transfer of a amino group "Transamination" through the exchange between a group of ketones in "Oxalaacetic", these three enzyme catalyzed reactions characterize the specificity of the protein part of the enzyme "Apoenzyme" and not to the enzyme assistant (co-enzyme), in the "Decarboxylation Tnansamination" the need for an enzyme and one assistant "Pyridoxal Phosphate".

Biology

The concept of this multi-genetic engineering and technologies based on splitting the DNA. By special enzymes called in specific engineering sites and connecting pieces formed with DNA from other sources. It is then formed on the multiplication of hybrid vital carrier capable of reproduction Cloning including bacteria and viruses have been used for the development of genetic engineering and other techniques electric and radiographic self.

This technology can be used for the preparation of the private DNA sensors for the purpose of search for specific genes or specific parts of the DNA as this technology can also clone parts of the DNA in large quantities by the private use of the enzyme reactant sequential PCR.

Genetics

Mendel discovered the basic principles of heredity and passed him a head later by scientists. As he concluded after mating successive generations of peas plants, that split successor inherits the characteristics of, according to a mathematical formula that could be the secret to life and then called the laws of Mendel. Then genetics born at the beginning of the twentieth century after the principles of, which were designated as Mendelian inheritance. Followed by several changes altered the traditional characters of life science to prolong and settled genetics then rolled up on the basis of discoveries that quoted life science version of the conventional version of the description and classification.

The Darwinian concept, was retreated which relies on the theory of evolution that the changes and the characters that we can get as a result of practice or reaction with the environment can be transferred by inheritance to the descendants. The hypothesis of evolution of species has been the trend due to the mutations, some aspects of the sudden change in the number of cases that called for the assumption that the diversity of animal emerged from mutations, and some of these changes may have inherited.

After that, the transfer life science version of the traditional formula and tradition description, classification, manifestations and modalities of organic evolution and the cell in its entirety to the life science microscopy, which focuses on exploration of the nucleus molecules and chemical structures. It is a mass of spherical material or oval that looks heavier than around it. Then it emerged that in the nucleus, clusters of fine particulate organic form, renamed chromosome or chromosomes that contain the genetic factors mentioned by Mendel each factor was called gene. The cell contain genes and for each type of species there is a special number of chromosomes in every cell of the human body, forty-six (46) chromosomes except in the female reproductive cell (egg) and sperm in the male sperm, each containing two (23) chromosomes. But there is one chromosome in a set of chromosomes of the male sperm determines sex of the fetus generated from the fertilized egg, it may be x or y. Not only the impact of these chromosomes to determine the sex of newborn, but also that genes determine the hereditary characteristics of male and female, then was found that the nucleic acids present in the nucleus of cells issued instructions for their growth and break apart and there are two types, DNA, RNA. Crick and Watson (1962) managed to develop an acceptable model for this structure which is composed of two bands of nucleic acid units (with four bases adenine, thymine, guanine, cytosine) in a corresponding sequential arrangement to RNA and such a model and a specific genetic information is transferred to RNA which controls the composition of proteins.

Biotechnology

Biotechnology caused enormous developments human, including those which were aimed at human use:
- Industrial electronics.
- Products of biotechnology space.
- Environmental treatment
- Extracting of oil by microorganism
- Medicinal plants.

The transfer of chromosomes from one cell to another is of chemical and living concept for the process of cloning, and its backbone. The genetic chromosomal transfer is not considered new, it has been exercised during the implantation of an egg in the uterus of a female in (1978) (the birth of a tube child - Louise).

Specifications and features of the cloning by the cloning is characterized with specific characteristics of chemical features of some new and some old concepts. These specifications are as follows:
- Converting an adult cell (totally grown) to a cell could reproduce without vaccination.
- Converting the reproducing cell to full a creature breeders (replica of the mother).
- Transfer of mature cell grown to fully grown live immature egg in the uterus of another object (the sheep, for example).
- Development of a new principle called the principle of Wilmurt can be expressed as follows: (very specialized mature cell from an animal that awakens static sophisticated genetic information in the chromosome to become a source of a whole new creature).

- This process of reproduction is not a sexual in the traditional sense (without sexual contact between male and female).
- The genetical cell can not be used through challenge and breeding.
- The cloning from adult stem cells facilitate the researcher to wait to see the nature of the thing itself before proceeding to reproduction.

Features applied to clones (the experiment of Ian's death)
- Converting a cell of the nipple of the gland of sheep (full growth - extremely) to a cell with a capacity of breeding and without sexual relationship.
- Transfer the mature full cell to fully live the immature egg in the uterus of a sheep.
- The cloned sheep of inherited properties, from the donor mother (the birth of Dolly the sheep).
- The sheep (Dolly) is not the nascent daughter and her mother, can be its twin.
- The sheep of the cell prepared for a clone of the breast, containing all the genes necessary for a complete sheep.
-

Old and new cloning

The cloning technique, distributed according to principles and specifications on several areas, including gene cloning embryonic cloning and cloning by nucleus. Gene cloning refers to the production of similar genes similar to the original gene, and the best example of this, is the process that occurs naturally when giving birth to identical twins, as a result of one split of egg genes and its distribution into two cells identical to the growth of each to produce transferred uniform embryo. But when the separation of embryo's cells from each other then process is referred to by certain embryonic cloning, each cell is growing separately to produce an integrated organism. Researchers have successfully cloned monkeys and frogs from embryonic stem cells match similar to the original. It was easier to deal with embryonic stem cells because they do not have any discrimination that they were not yet have been developed to convert to and brain, muscle and other organs, in the wombs of their mothers. But it turned out with the progress of development, the property of discrimination occur a change in the DNA. The passivity of embryonic cloning researcher stresses the intrinsic character of being unaware of what will emerge.

The third cloning, which represents the third occurrence of mature cells, a researcher can wait to see the nature of the thing with his own eyes before they proceed on reproduction. This represents a new type of reproduction, which is using the nucleus of an adult cell, adopted by Wilmot the specialist knowledge of embryos, which represents technical innovation for the reproduction of fetal sheep from an adult cell, and he was able to clone sheep, genetically modified by human genes to produce factors that deal with clot the blood. The Dolly, of Wilmot considered as the first which was cloned with the latest technology and adult cell cloning technology has international new principles of life, especially relying on a somatic cell, and successful steps can be carried out each and every one has a recipe associated with these principles:

- Dealing with the cell donor (host) intended for cloning and extracted from the membrane cells of a pregnant sheep is characterized by white color and necessary genes needed to form the purpose complete sheep. This step represents a use an adult cell transferred to a cell with the ability to reproduce without sexual mating to become a creature without full sexual intercourse and instead of being subject to embryonic cell fusion.

- An unfertilized egg cell that has been stripped from another type of sheep with a black color in the laboratory and the nucleus was removed, but the cytoplasm remain in the proper position.

- The defined cell donors exposed to famine and by halting its development and to prevent divisions and food resources for a period of ten days and surrendered to a state of sleep. The egg were put near the nucleus of the cell donor and prepared for cloning. Using weak electric firing bursts then nucleus and the egg unite and start acting as fertilized egg.

- Allowing the embryonic cells to grow and divide and then formation of embryo in the pot laboratory, implants in the uterus of sheep with black head. The sheep that was born become and was named an international replica of the white sheep of the cell donor not the same as the black sheep used for the lap.

Bio communications of life

Communications carries out number of operations which include successive generations such bioelectrical- communication and chemical, reflects the continuity of generations, which represents the survival of reproduction in different ways. The mainstream way is connection in the process of sexual relationship resulting in the integration of the male cell (sperm) female cell (egg). Reproduction may occur without convergence between male and female, as in the cloning processes, which we have mentioned, or, as occurs in the primitive animals, with single cell in a simple dichotomy, as do animal and amibia, paramseyoum. In the process of sexual convergence the number of chromosomes go back to the original number in all of the cells of the human body and becomes a forty-six chromosomes (23 from the nucleus + 23 from the sperm). Therefore, the characteristics of many genes on the chromosomes are transferred. It is worth mentioning that the vaccination is not described in the same way it may be externally or internally in the fish and some other animals, the sperm and eggs put in the water in which they live. The artificial insemination is carried out by sperm from a male and put in the female's vagina to cause pregnancy for the purpose of the transfer of the qualities required to a large number of females.

The other type of communication named electrical contact is carried out via the nervous system, which includes the central nervous system (brain and nervous system or spinal cord) and peripheral nervous system (nerves of the brain, nerves and spinal cord) nervous system (self-sympathetic and para sympathetic). The brain, which represents the main component of the central nervous system, contains about 12 billion nerve cells that do not divide, and when they die does not change with other cells. The nerve cells for example, connect impulses and the installation of a specific architecture. It is worth noting that electric current flows in a given direction and one inside the neuron, and to understand how we entry into nerve impulses in the nerve cell should be conceived that in all cells there is a difference in voltage between the inner and outer surfaces of the cell membrane surrounding the cell. The chemical communication is carried out by glands that produce hormones by channels "Ductless Glands" then they carry out considerable influence on many of the functions of the body.

.

Later several techniques developed, including electron microscope, where it was possible to obtain precise details of the minutes. To reflect the elements of body composition other technique called X-ray have been used to study, the order of atoms in many of the biological compounds.

The isotope techniques have continued as a useful tool in the search for the secret of life is as well as a means of treatment. Researchers have manufactured hundreds of radioactive isotopes, generated from the non-radioactive elements in nature, including sodium, sulfur, calcium, chlorine, copper, cobalt, gold, iron, mercury and silver. The main uses of these element the process of photosynthesis, as well as follow- up to 14C in the development of a new technology called radio- immunoassay which can be used to determine the concentrations of compounds found in very small quantities, especially hormones.

Used several techniques to separate, including electrical and deportation chromatography. The first has been used to isolate many of the vehicles life and purification, and the first uses and which still constitute the mainstay of the first structural study of structure of proteins and amino acids, thanks to the use of positive ion equivalent.

Genetic Engineering

Genetic engineering caused major developments in life sciences, including applications in medicine, which includes diagnosis and treatment.

The concept of the genetic engineering and technologies based on multi-splitting of DNA By special enzymes work on specific sites and then linking the pieces formed with DNA from other sources, is then the proliferation of hybrid is able to reproduce on vehicle "Cloning", including bacteria and viruses which have been used to develop genetic engineering techniques and other such as electrophoresis and auto radiography.

This technology can be used for the preparation of special DNA sensors for the purpose of searching for specific genes or specific parts of the DNA technology to clone parts of the DNA in large quantities by the use of the enzyme "PCR".

Applications of genetic engineering:

- Gene therapy: It is used in the treatment of some diseases, including the treatment of brain tumors and in the reduction of cholesterol in the blood.
- Genetic mapping in humans: one can look at the matter in the future after the location of many human genes.
- Stimulation the immune system to produce antibodies more efficient and accurate (a vaccine against the virus of hepatitis "B").
- Production of proteins of medical importance.
- Determination of the nature and location of genes for some genetic

Genetic engineering has brought tremendous developments in the life sciences, including applications in medical science, which includes diagnosis and treatment. The old and the new in cloning

Cloning technology is distributed in accordance with criteria and specifications on several areas, including gene cloning and gene cloning and reproduction nucleus. Gene Cloning refers to the production of identical genes are similar to the original gene, and the best example of this process is what happens naturally when you give birth to twins as a result of a split similar genes per egg and distribution identical, for the growth of each of them to produce each one embryo transferred uniform. But when you separate the fetal cells from each other in a certain phase is called a gene cloning, where they grow each cell separately to produce an integrated organism.

Researchers have successfully cloned monkeys and frogs from the cloned gene cells match similar to the original, and is believed to be easier to deal with embryonic stem cells because they do not possess it any they were not been exposed after the convertible evolved into hair and brain and muscles and other organs. It is in the wombs of their mothers, but it turned out with the progress of development of the property discrimination occur a shift in the DNA that negative genetic cloning researcher emphasizes the fundamental recipe is being not knowing what will emerge from it.

The third cloning which represents the occurrence of the completed cell growth, the researcher can wait to see the nature of the thing with his eyes before baptizing a reproducible. That this type represents a new cloning and that is using the nucleus of an adult cell and adopted by the world Wilmot specialist knowledge of the embryos, which represents innovation embryonic technique to clone a sheep from an adult cell, is also able to clone sheep genetically modified human genes.

Mathematics

Emphasis on the unity of mathematics and philosophy, includes several directions, such as, applied mathematics, engineering, life science, mathematical economy, financial mathematics, probability and statistics, operations research, mathematical research and other applications have also been reaffirmed through mathematics public programs, In particular, special importance to the modeling of mathematics, has gone some universities to develop a draft "project" and mathematical clinic confirm this applications. Moreover, studies have developed interface such as mathematics statistics, mathematics philosophy, mathematics and the economy, mathematics and physics, mathematics and engineering, mathematics and computers, and others.

Computers have been introduced as a mean to teach some subjects of mathematics and devise new methods of teaching an alternative to the currently prevailing lecture and attention to the employment of culture and mathematical awareness.

The unity of mathematics as a philosophy and the name of mathematical sciences include many aspects such as applied mathematics (engineering) and life sciences, mathematical economy, financial mathematics, probability and statistics, operations research, mathematics research, and others. It was also stressed the applications of mathematics through public programs, and in particular to give special significance to the mathematical modeling, has gone some universities to develop a draft "Project" clinics "Mathematical Clinic". Moreover, studies have developed an interface such as mathematics and statistics, mathematics and philosophy, mathematics and economics, mathematics and physics, mathematics and engineering, mathematics and computers, and others. Computers were introduced as a means to teach some math and attention to promotion of culture and awareness of mathematical mathematician. To provide students with the qualifications can be expressed as follows:

- A sufficient culture to qualify them to understand the difficulties.

- Encouragement self- confidence of all to address the difficult technical issues.

- The amount of uncertainty makes them ask the right questions.

- The amount of persistence makes them continue to search for appropriate responses.

- Discretion to choose what is right.

- is worth mentioning that it was the evolution of mathematics inherent in physics, then ripped mathematics and liberation and began the development of mathematical systems abstract, using set theory and mathematical logic, and this, use the workers and researchers in mathematics, and ways to create a broader understanding of the economy and population change, earthquakes and human behavior and medicine, introduced as a result of the impact of new terms of reference including sports economics, mathematical and theoretical biology and science of sports ground and sports psychology and linguistics sports and the adoption of the so-called mathematical modeling introduced then "applied mathematics" the jurisdiction of which also included mathematics space, mathematician energy, either pure mathematics such as algebra substitution and the theory of the symbol and ring theory and the theory of numbers topology and algebraic and differential topology and algebraic geometry and calculus equations in differential and integral intervened as a platform to build models. Grow mathematics varying speeds and expanding their uses and trends in the areas of knowledge, including the construction of mathematical models, and cooperation with the flag

Mathematics contributed to the evolution of technology such as manufacturing computer with very high computational capabilities, the specialized topics bases in Computer Science, Mathematics robot, mathematics and artificial intelligence, and engineering computing, that the impact of computers not only on mathematics developed (computational science) and the emergence of computational mathematics.,

The mathematics concerned with the study of common properties for sports models and the development of algorithms, but has spread to other sciences.

Computer Science and Informatics

Recent trends suggest that the evolution of Computer Science from the computer discovery and manufacturing in the early sixties. Computer science was not then separated itself along the lines of pure or applied science, but they grew up in the embrace of scientific departments, especially departments of mathematics and electrical engineering departments.

The computer hardware and curriculum were not possible to sufficiently and necessary for the different compartments and focused the first two- way computer materials is the direction of engineering ties to computer components and architecture, and the second trend is the compiler. Number of programming languages in science and engineering faculties used to solve mathematical problems and mathematical, engineering or for the purposes of modeling and simulation. And the evolution of computers in terms of physical hardware "Hardware" and software operational "Software" to develop computer science as one of the basic science, where it was subjected to the axioms of this science and the philosophy of pure science and applied at the same time. Since the late sixties, introduced independent scientific sections of computer science "Computer Science" or computer science and information "Computer and Information Science" or computer education "Computer Education" or "Cybernetics" or processing of information "Data Processing", or under other refers to one type of information processing.

Development of computer education even intervened materials with a large number of scientific and humanitarian and computer hardware has become part of the curriculum requirements of the various articles of pure science, applied science and humanities. As a result of the marriage of computer science with other fields of science and new scientific knowledge, especially in the application, example such as field of computer linguistics, "Computational Linguistics" as a result of overlap with computer science, linguistics, or a discipline called decision support systems "Decision Support Systems"

The evolution of computer sciences was started since the discovery of computer laboratories and manufacturing in the number of American universities and the British in the early sixties.

The computer science was not as separate discipline as pure science or applied, but had its inception in the arms of scientific sections, especially sections of mathematics and electrical engineering departments.

As computer hardware and curriculum were not enough to be in different department at the beginning. The first computer hardware trend was related to engineering and architectural computer components. The second trend was the software programs, which examines a number of programming languages in science and engineering faculties used to solve the issues of mathematics computational and engineering and statistical purposes or modeling and simulation. The evolution of computers in terms of physical device "Hardware" and "Software" began to widen in scientific, theoretical and practical trends. This expansion has led to accelerated development of computer science as one of the basic sciences.

The large number of universities and colleges established the formation of separate departments for the computer science, such as, computer science, information computer, information science, computer education or cybernetics, or information processing, data processing. The resulting marriage of computer science with other fields of science, new scientific knowledge was created, especially in the field of application.

For example, the evolution of computer "Computational Linguistics" as a result of overlapping computer science with the knowledge of language or field of knowledge called "Decision Support Systems" overlap between computer science or science information systems and management science and operations research and statistics.

Computer Science evolved since the discovery of the computer and processing laboratories in a number of US and British universities in the early sixties. Computer science was not at that time have crystallized a separate note itself, similar to science pure or applied, but grew up in the embrace of scientific departments, especially math and departments of electrical engineering departments, where the computer materials and methods were not possible to needed to do different sections and sufficiently, also focused Computing material in two directions one of them has to do with engineering and architectural components of computer software and the other direction, where he teaches a number of programming languages in colleges of science and engineering, and are used to solve mathematical and computational, engineering and statistical matters, or for the purposes of modeling and simulation.

And the evolution of computers in terms of physical devices (Hardware) and operational software (Software) began expanding scientific and theoretical material and scientific and this developer and the rapid expansion has led to develop computer science as one of the basic sciences, and a large number of universities and colleges to form separate sections to computer Computer Sscience or science computer and information computer and information Science or computer Education , Cybernetics or information processing Data processing or under other headings refer to one type of information processing. The expansion of computer education materials so that overlapped with a large number of scientific and humanitarian materials .

Computer Science has developed since the discovery of the computer and processing laboratories in a number of US and British universities in the early sixties. Computer science was not at that time have crystallized a separate note itself, similar to science Pure or Applied, but it was growing up in the arms of scientific sections, especially math and departments of electrical engineering departments,.

where the computer materials and methods were not possible to needed to do different sections and sufficiently, also focused Computing materials in two directions One of them has to do with geometric and architectural components of computer software and the other direction, where he teaches a number of programming languages in science and engineering faculties, and used to solve mathematical and computational, engineering .

The expansion of computer education even overlapped articles with a large number of scientific and humanitarian materials and computer materials become part of the curriculum requirements of the various Pure and Applied Sciences and humanitarian materials.

This has resulted in computer science with other sciences knowledge and scientific fields and new, especially in the field of application for example, the development of computational linguistics field of Computational Linguistics as a result of overlap so computer with linguistics or cognitive field between computer science or information science systems and science Management and Operations Research .

Virtual Informational Technology (IT) Knowledge

Virtual community association is characterized by information and the presence of different types of interactions knowledge, information and economic and the presence of e-government.

Therefore, virtual community knowledge society replaces the real values and standards of a new hypothesis occupies a distinct position in the knowledge society assumption; represent a model of high proficiency, difficult to isolate it from reality. The virtual universities are academic institutions aimed at providing levels of high-quality education for students in their residence places through the world wide web of the internet where these universities work to create an electronic educational structure, meaning that there is no need for classrooms or buildings or student rallies in classrooms or exams. In other words, these universities established ranks of virtual default continue to conduct tests on after the adoption of advanced programs and therefore it requires Iraq to take the experience of virtual universities in view of what can be provided by funds, where low-cost student and regular pressure on universities. The virtual universities create platforms for a variety of electronic and administrative services and create an oasis of virtual science.

It is noteworthy, the idea that emergence of virtual universities depends on the use of e-learning in terms of curriculum, teaching methods and techniques according to the explanatory educational academic standards developed and invested in self-education "self learning" effective and to obtain academic degree.

Then It began the development of computer hardware and adoption of operational programs "Abermjiat" This rapid development has led to the development of computer science in a large number of universities and colleges to form a separate fields of computer science, such as information, education and computer cybernetics , information processing and data processing interfere with a lot of material.

Materials have led to this mating with different fields of knowledge, especially in the field of development and field applications, for example, "Computational Linguistics" as a result of overlapping of computer science with linguistics.

The most important disciplines in the near future, computer technology and robotics, software and information systems and systems engineering and engineering software and information technology level, and information systems and neural networks and virtual reality and networks of fiber optic and information security, encryption, and data processing and systems of teaching and learning systems design, production and neural networks and systems knowledge.

The Iraqi universities new specializations in the field of computer, such as computer technology and robotics, systems engineering, artificial intelligence, information management and computer security, information systems, and virtual reality, and decision support systems, machine translation neural networks, and the protection of information networks and knowledge of computing and information systems.

The foundations of bioinformatics

Different forms use the methods and tools for a variety of information consistent with the idea of the basic units of DNA genes, but it is at the same time vary depending on the sequence of these same units and the extent of its simplicity or complexity. That the circumstances that prevailed in the incidence of various diseases, and related to these conditions of major change which is to a large extent in increasing trend towards updating of information about human biology, including abnormalities and diseases related example is characterized by the fact that the cell individual contain a third version, it is determined by microscopic examination.

The tremendous progress has been possible to follow changes in the DNA including mutations responsible for many genetic diseases. Importantly, however, that recent data is heavily dependent on modern technologies that are meant connected to the researchers of these operations and become one of the fundamentals of bioinformatics of science-related advanced and effective device, hence the importance and seriousness of this means.

A common relations between computer scientists and molecular biology scientists can do to provide the computer world that rules necessary to encrypt the D.n.a and thus progress in the future and are shown adequate important information then to determine the organism interpretation.

Moreover, there is a possibility for the development of the computer using the D.n.a and the use of mathematical rules as symbols When adding a certain amount of the D.n.a all the information in the computer Storage. The idea of a computer that appeared in the 1995 adaptation of the DNA afford to process information.

Finally, the biology is no longer limited to the professionals but entered into computer science and scientists who are beginning to learn this science and participate in the research teams of specialists. Computer life develop, and can be imagined led by computer scientists and biology became note in terms of information which is called bioinformatics and life size computer components billions of times smaller than silicon chip that is characterized by a very huge storage capacity.

From a computer look the DNA and installation of structural represents a system intelligent for the storage of information and computer scientists have become accustomed to deal with the digital binary system to express the alphabet and numbers, symbols and diagnosed directly alphabetical character quadruple in the DNA in order to encrypt messages and that all three-relay in the DNA

That the information stored in sophisticated computers evoke different senses dramatically, as actually live researcher within the computer and sees human life is stored and the whole picture of technology modern era necessarily means exploit and develop as the many researchers in the countries that have contributed to the success of the Human Genome Project that developed means and comparisons.

The Human Genome Project (HGP) Applied tool of information for scientists and researchers with vital information on the basis of the genes is crucial to the success of research programs for the purpose of basic human life according to information identity.

The Human Genome Initiative was the first time in 1988 and aimed to find sites about (100.000) gene in the human genome in D.n.a and expresses the 24 pairs of chromosomes, turns into an information content when follow the sequence of rules need to be based on solutions of Computer science , mathematics , statistics and empirical science.

Note that the Computer Science offers mostly contributions to programs and solutions are characterized by the skills led to the invention of access in programming languages and describe the information that implements a specific order and provides methods to describe the complex dynamic processes with a number of software Lines.

After researchers and observers have said that -first century will be the century of biology and analytical power resulting from (HGP) that will be interpreted radically and medical research as it has been:

Finding the nature of genomes and the nature of its composition and organization in different scientific institutions.

Acceleration in the implementation of the project, which plans to accomplish during the 15 years, but the technical progress shortened the time to ten years.

• Finding in genome content to know the type of information or material contained in contact to diagnose all the genes the (100.000) in the DNA of human as well as determine the sequences of nearly three billion chemical bases.

Study the nature of the information stored in the evolution of computer technology and elaborately efficient sequence and the evolution of the tools that contribute to the analysis of information.

Progress and development in Pure Science, for example, led to the development topics and disciplines and new disciplines and sciences Interfaces new were not known during the first half of the last century. The results of such events and large-scale changes in the structures of study and research approaches, as these developments in the developed world universities turned into curricula study and research.

Although modern science bioinformatics very aware of the complexity derives its assets and the accountability of non-science, the most important of mathematics and computer science as well as many of the explanations of medical science resulting from human genome studies (gene content) for different objects, and as well as the use of electronic information to understand Genetic network and the development.

Learning

The e-learning is intended to have several different types of education and training is different in terms of the nature of the educational process and content, methodology and fundamental difference between education and tele-education.

It is activated in this regard by the university of Baghdad programme of lectures by visiting Professors through e-education system which is being built in collaboration with the University of e-learning in Canada, where these techniques will open up prospects in the wider access to scientific developments at the international level.

The Quality Assurance and international partnership with Open University are applied by the Arab Open University. The most prominent features of this university are to lift restrictions on the admission and registration, contrary to what happens in most universities of open education in the world. The education system in the Arab Open University bounds student to spend time in mandatory meetings and seminars. This Open University believes whole importance of quality assurance and seeks first to measure the performance quality university and then make the means of quality support and assurance. It has adopted measurement tools and has evaluated the quality of British Open University and divided the university system to ensure quality and measurable ways and means of support.

The Arab and international experiences in the field of e-learning at the university began in Education in 2000. Characterized as an important educational issue in our contemporary terms by giving to science students the ability to search and investigation and find modern information. It also confirms that the scientists learning does not mean merely exploiting the potential of modern technology in the delivery and knowledge. Students can now attend the university located in places other than those that live in. accordingly E-learning includes concepts and philosophy, objectives, certain patterns, pillars, requirements and mechanisms.

Distance Education

In order to shed light on distance education and its importance, it is suggested the following points:
- The reality of distance education in terms of goals and ambition.
- The importance of the challenges facing the movement of distance education.

- Effectiveness of distance education compared to the traditional model of education.

It is also stresses that the Arab strategy for distance education in 2004 is to:

- Liberate human beings rights and maximize its contributions to progress of development.
- Meet the need for integration with other learning systems, including the education sector.

The distance education system is an educational model based on modern technology, but there are many obstacles facing the problem of application, such as, recognition, criteria, measurement, standards and quality and others. The positive results on the importance of distance learning as model of education compared to other models, suggest thinking in different forms and regulations and renewable methods of distance education to suit the point in time of the third millennium.

Adoption systems of distance education through computer networks on the concept of the overall approach include a series of general educational measures in electronic form so as to be accessible to scholars. Distance education is one of the important means of communications and technological revolution in the transfer of knowledge and its uses to develop and employ them in the development of human capabilities and enabling environment for communication to the world of technology and information between individuals and among all sources of knowledge everywhere reach these networks and produces the network of learner direct contact with science continuously. There are also regularly available information, photos, and recordings through the web, along with holding meetings and symposia.

Education Using the Internet

The use of the Internet has become a reality impose itself as a modern method used in university education, but there are many things to identify clearly the benefit from its use and encourage its spread in future, including:

- Descriptions of output achieved by the kind of education.
- Clarification the methods of coordination between various education systems through the network.
- Development of assessment methods appropriate for students.
- Update instructions in line with the software used in line with the software used for this type of education.
- Cooperation colleges in the corresponding decisions similar to this type of education.
- License modernization so as to protect the rights of electronic publications.
- Determining academic standards for admission.

Activated Education

The activated education addressed to stimulate skills, to attach a variety of potential users, to remain with them throughout their lives, to take responsibility for themselves and their learning. It is evident that such constructive learning environments through successful players should be well prepared containing different process and the link between qualified teachers with a renewed mind movable teacher has features of mobility.

Nano

nanotechnology, the technology that deals nanometer scale (metric unit of measurement), which is manipulating atoms for the manufacture of automatic equipment and information does not extend far a handful of atoms, and then anything can be made by micro- physics is quite different. The first part of the term to reflect the unit of measure (nano= 190^9 meters)

the term nanotechnology in Engines of Creation written in 1986 and said the possibility of seeing the future of the armies of machinery hidden carry oxygen and nutrients and waste, and manufacture of atomic- sized machines called complexes pads that hold individual atoms. It was found that the control of the maize one and move freely and easily from attributes of nanotechnology.

This technique showed high density in the form of recent innovations in many of the global scientific publications. Including the mutations responsible for many genetic diseases and therefore provide in the future and relevance of information is indicated task to determine the organism is an information system for the manufacture of proteins and other compounds in spite of the inability to resolve the issue of structural triangular shape of proteins, despite the existence of mathematical models serve the purpose.

A fruitful field of nanotechnology research in many parts of the world and in government labs, commercial and academic, have emerged, according to the products on this technique such as sewing pants of fiber and manufacture of precision tennis balls retain flexibility. In the near future, computers will appear smaller tubes made of carbon atom chips represent the atomic scale wires and high strength to build elevator to space, and plants that will manufacture computers minute integrated directly with the human brain to increase intelligence. It is clear from the examples mentioned that nanotechnology is a technology that will change very little minutes every aspect of human life and giving people the ability to control the material, and this technique is the most important applications of medical treatment of human beings through the introduction of precision instruments within the cells to repair infected objects from within or for the diagnosis of patients as well as some developments on the mechanisms of control cells. The first medical use of this technology has been developed a device implanted in the body that may manufacture.

The energy comes from bio-fuels in the cell and the purpose of this engine is to integrate machines in living systems fully, and use some of the scholars of this technology to produce nano-bombs to kill cancer cells. A team of other customers of this technology to produce nano- bombs to kill cancer cells. A team of scientists of any other industry, the crew of siliceous teeth not larger than the size of the cell that can swallow the red blood cells and re-launched into the bloodstream, either antibiotics nano-particles "Nano biotics", which new types of antibiotics contribute to solving the problem of resistance of some types of bacteria to drugs, as well as the modified bacteria organisms, are converging nano- tube micro- rings 2.5 nm diameter amino acids and small hole walls of bacteria are infectious.

Researchers believe that the future of medicine is moving towards nanotechnology that will change medicine, as future devices that will work within the human body to diagnose many diseases and treatment. Russian scientists in the field of quantitative light and laser physics to reach a new discovery has been called the needles agency, a new type of X- ray beam, or special characteristics, as containing the elements of nano- any electronic material on subatomic particles that do not exceed the measurements of nanometer dimensions . It has also developed the first computer chip companies auction that could contribute to increase the power of computers and a reduction, while reducing the amount of energy consumed by the chip is composed of cylindrical molecules of carbon atoms in diameter than a billion to a part of the linker carbon (smaller than a hair a hundred thousand times).

Hot issues

These issues raised by biotechnology as a result they have become at the forefront of basic research and applied consistently reached new levels of progress and complexity and can be far- reaching impact and positions that require scientific, political, moral and social. These issues vary with varying impact and could be referred to some of them:

- Cloning.
- Human Genome.
- Gene therapy.
- Map of the protein.
- Food and genetically modified organisms.
- Advanced technologies (nanotechnology).
- Vital information.
- Monoclonal

antibodiesBiotechnology, medicine, agriculture.

- The discovery of disease- causing genes.
- Forensic – DNA, Fingerprinting.

- Biotechnology and biosafety.
- Biotechnology and environmental balance.
- Scientific strategies of biotechnology.
- Research and development and stops future.
- The role of education and training in biotechnology development.
- The twenty-first century is the century of Biotechnology vitality and prospects and challenges.
- Technical aspects of genetic engineering.
- Bioinformatics

It can focus on hot issues of the following:
- Cloning.
- Human genome.
- Gene therapy.
- Genetically modified food.
- Ethics.
- Genetic engineering and the internet.
- Biological weapons.

Cloning

Cloning techniques are distributed according to principles and specifications of the theory and practical on several areas, genetic cloning and reproduction by nucleus. Gene cloning refers to the production of similar genes resembling the original, and the best example of this when giving birth to identical twins after a split genes of one egg and its distribution into two similar cells and grow each of them to produce a separate identical fetus.

But when the cell of the embryo is sperated in the process is referred to by a particular genetic cloning, and it is growing each cell separately to produce an integrated organism researchers have successfully cloned monkeys and frogs of the cloned embryonic stem cells, similar to the original match, and it believes that it is easy to deal with embryonic stem cells because they are not discriminated by the (were not an evolutionary has turned to hair and brain and muscles and other organs) is in the wombs of their mothers, the negative aspects of cloning, the genetic test will be based in is not what will the world of him.

The third cloning occurs from mature cells, the researcher can wait to see the nature of the thing with his own eyes before they proceed on reproduction. This type of cloning is new and are using the nucleus of an

adult cell adopted by Wilmot (embryologist) to clone a sheep from an adult cell, as well as it has managed to clone sheep, genetically modified with genes to produce human factors and blood clot of the most famous that carry human genes. Dolly, the most famous reproduced onganism that carry human genies while the cloned Dolly the first to reproduce the new technology. International principles for the reproduction of new life based on the cloning of a somatic cell is a mature steps each one of them have the status associated with these principles and is the following:

• The use of a cell donor (host) intended for reproduction of cells extracted from the membrane to view white pregnant sheep featuring the genes needed to form a complete sheep for the purpose of conversion to a cell capable of regeneration without preproduction and become a creature without full sexual intercourse (without vaccination) and instead of being cell embryonic viable fusion.

• Using an egg cell is extracted from sheep attached to another type of black- headed in the pot tester removed while retaining the nucleus with cytoplasm to put it right.

• The cell donors to famine and a halt to development and divisions, and preventing food resources for a period of ten days and give in to the state of sleep.

• The removed egg from the receptor cell near the nucleus of the cell donor prepared for cloning by using electric induction, small electrical firing bursts per nucleus and the egg unite and begin to act as fertilize.

• Allowing the egg genetic cells for growth and division and the formation embryo in the laboratory, implants in the uterus of sheep with black head and become the sheep that was born and was named an international replica of the white sheep of the cell donor and not the same as that used for the black bosom.

Genome

The human chromosomes is 46 consisting of strips of the double helical DNA wrapped circumvent complex shapes of helix, normal and high and consists of the DNA with four units of high repeated synthetic (nucleotides incomplete oxygen), each of which consists of three components: nitrogenous base and sugar phosphate penta and not organic.

There are four types of nitrogenous bases and symbolized by TCGA arranged in pairs along the stretch of DNA and the numbers of secondary units in the DNA molecule. Approximately 3×10^7 base pairs in each of the cells of the human body, and the length of DNA equal 8×10^3 times the distance between the earth and the moon and bigger than the distance between the earth and the sun 300 times (the length of all the DNA, In the body 2×10^{10} km).

The gene (one gene) constitutes a piece of DNA it consists of a large number of secondary units and in the molecular weight of the gene is 600×10^3 and is a very long string of four characters, and each character represents a nitrogenous base. There are usually genes in the nucleus of the cell and molecular genetic consists of two bands are linked together by special bonds, each other on some twisting spiral and there is peace on the same wrapped that consists of a sequence of nitrogenous bases, or nucleotides that contain the bases arranged in a manner different from the gene to another and then discriminate organism from the other because of all the genes governing cell functions, guidance on ways making a specific protein or another compound with medical importance, hence a single gene responsible for the general one recipe and therefore we find that the qualities beauty, shapes and colors that each one of them result of a single gene or the number of genes. Genes are transmitted from parents to offspring by mating the structural change in terms of affected and then the subsequent processes of making many of the compound causing the disease, which may be cancerous or always defect organisms.

The genome is intended to aggregate the DNA (genes) of the bacterial cell example, contain about 200-300 gene, while the genome of a human cell include the thousand times as much as the genes found in bacterial cell 200.000 to 300.000 and the organization of these genes depends on the number, in the chromosomes of the cell Eukaryotic (human cell) is more complex than the primitive cell nucleus (bacteria).

The genetic map represents the order of genes (genes) within the cell chromosomes and that this arrangement within the human chromosomes is more complicated than other organisms. Thus, the process of discovering how to arrange these genes, given the sheer number and complexity associated with variation built and responsibility to control complex cellular functions. Hence the decoding process and diagnosis and scheduling of full human genome by genetic map and the preliminary draft of a preliminary genetic blueprint of human genes and the previous process is equivalent to a significant scientific breakthrough, scientific achievements made during the twentieth century, including the discovery of penicillin and landing on the lunar surface and use a computer and other discoveries. And according to this perception announced 26.6.2000, the end of the main phase of the Human Genome Project, which

represents the first achievement in the twenty-century atheist and the development of the draft map is almost complete and a preliminary blueprint for a human gene content of the human genome and was named the human genome. It had been prepared jointly by both the research centers m the United States, Britain, Japan, France, Germany, China and other countries with long experience in genetics and genetic engineering, funded by 18 countries.

This map is characterized thoroughly without gaps up 99.9% and 97% of the components of the human genome, has been decoded and 85% of the sequence and gene order has been tabulated and analyzed. The rest of the map requires additional time to accomplish. The map we have opened a new era of molecular deal with the situation of life, has made it clear that in a manner distinct and different, Dr. Ahmed Zewail Nobel Laureate in nanotechnology, said that the molecules are arranged genes that act in the movement of one can not detect them, but that can be pursued with sophisticated and sensitive to femto second.

The discovery of the human genome and complete the approximate locations and sequencing of this large number of genes input to future developments are to:

- A new look for the human body.
- To find new ways of treating diseases (gene therapy developed) such as AIDS, cancer and heart disease.
- Correcting the genetic errors.
- Organ transplantation.
- To address the social and ethical consequences. Sustaining life.

The battle against cancer, AIDS and other incurable diseases are ready to locate the responsible genes for these diseases that have discovered the map and according to that can solve the problems of treating these chronic diseases and decrypt secrets, and the negative aspects of this discovery is immoral exploitation of the future such as racial discrimination in accordance with the genetic composition and control of human qualities and it requires the issuance of special legislation related

to human rights and prevent the future destruction of rights (death of persons with disabilities and life of the insane, for example).

Gene therapy

Thousands of diseases due to the presence of genes responsible for the appearance and many of them dangerous to humans and non- treatment or cure and concerns and applications of biotechnology to find what is known as gene therapy, which is either by bringing the damaged gene or gene intact repair defective gene. This could be done through the intervention to repair the gene in somatic cells, or by intervening in the cell construction.

In order to spread the use of gene therapy it must be certain that the expiry date and free from damage and researchers that must be able to transport techniques and control gene expression in the correct and consistent, and should not obscure the international success of the many risks carried by this treatment.

• The genetic balance of any human being is the only thing that can not be replaced but must be preserved and transferred to the generations of while it is possible.

• Here we must make sure that it can allow gene therapy in somatic cells, and prevention must be in the manipulation because of its many negative consequences both in terms of genetic or moral.

• Gene therapy in somatic cell only affects the individual patient treat him, while affecting.

• Gene therapy in cells on the construction of successive generations.

Recent progress in the field of gene therapy

It has become possible to do some practice in the field of genetic medicine with the evolution of technology DNA "Recombinant DNA" as has been addressed most of the problems related to the production and disposal of genes and to consider their ability to modify objects, and the laboratory tests on animals proved that non- genetic, genetic medicine can be successful. It has been treated several human cells in inherited tissue laboratory for the use of retroviral vector. And proved the possibility of a peace process, the introduction of white blood that had been genetically engineered in the patient. It was also carried out several recent clinical tests for human genetic medicine, culminating in successfully treating a patient in the loss of immune complex advanced emergency resulting from the adenosine.

It seems that gene therapy is the only way to cure genetic diseases or chronic (such as cancer, Acquired Immune Deficiency Syndrome). In this case, there is an objective one is to improve the health status or save the life of a patient work of the highly desirable, and then there is no difference between the unit body and the unity of the genes. Gene therapy holds great danger is the use of this technology in order to improve the human race; can we change the balance of genetic risk for the human species. This may have dire consequences especially on the reduction of formal diversity. This method, which promises also carry the hopes of many fears. Do we have verified that all the risks in the long run, where would we be? What are our borders? Where the border between the correction of what eugenics? The achievement of this modern way of thinking must be accompanied by a profound and moral debate.

General characteristics of the human gene therapy

It was discovered more than two thousand genetic disease; all affect the genetic information in the patient and move it to the next generation. In order to restore the natural functions of the victim there are two ways in gene therapy.

- Gene therapy in somatic cell.
- Gene therapy in the cell construction.

In each of these methods a special set of scientific and ethical considerations. Construction cells are sperm cells and egg cells, which include the rest of the cells in vivo somatic cells. The gene therapy of somatic cells in the introduction of DNA in this type of cells so the added gene in the progeny of the patient on the contrary, affects gene therapy in cells on the construction rights in the early stages of embryo development.

Gene therapy in somatic cell

Before embarking on any attempt of human gene therapy it must first determine the exact mutation that leads to a particular disease. This information is not available currently, but in terms of a small number of diseases, but current advances in genetic engineering point of what will happen soon on detailed analysis of the genetic and second it should be identified on the type of mutation affected the cell in the body and their genetic transformation. Only current clinical trials to treat somatic cells, which are based on the introduction of a gene in somatic cells of a small child or a young man. Thus, cells not exposed to structural change, which prevents the transmission of the gene to offspring and this, is something which makes having a person who benefited from the treatment is always vulnerable to this disease. The structural gene therapy did not apply to humans, as it was rejected on moral grounds. But a team of researchers are currently considering the possibility of its application in the treatment of incurable genetic disease and it comes in all cases. The treatment depends on the addendum, any patient that the gene dysfunction and genetic causes of the disease will not heal or replaced, but added to the cell intact copy of it and this method of treatment do not apply except in cases of genetic diseases caused by genes elected. In general, gene expression is not valid only in a given tissue can be determined in different ways in experiments conducted on the living body, determines the quality of the input method the target tissue. Incomes through the trachea transfer of genes into the pulmonary epithelium, and injection in the liver gene transfer to liver tissue, either tumor injection in the objects is transferred genes into tumor cells also contributes to the carrier of the virus in determining the target tissue.

Cells contribute to construction in the genetic heritage of successive generations. Gene therapy through the cell affect the construction of genetic stock to his descendants too, and then the sum of the genetic traits of humanity. The majority of scientists that may not be morally any attempts of this kind of treatment, another group believes that the gene therapy in the cell construction is the only way to eliminate genetic diseases suffered by millions of people.

Different methods of gene transfer

- Viruses: Retro virus, including the only to ensure that the genes transmitted via cellular divisions. We therefore consider these viruses are the most successful means of gene transfer in the laboratory, where they allow in principle a final treatment of genetic diseases.

- Chemical methods: There are numerous studies on the possible use of fat bodies and compounds multi positively charged.

- Physical methods: The target cells is ejected shells are small technisten DNA speed due to electrical discharge or explosion compressed gas.

- Intramuscular injection: Move DNA the intentions of cells around the injection site but does not merge with it, but stay for extending from a few weeks to a few months to form ring.

To deliver a specific gene or gene fragment to a cell, you should test the appropriate carrier depending on the cell type and the type of genetic defect, but for the gene replacement method, i.e. a shift in the form of direct mutated gene in position, it can not be used because most of the known vector vaccination consecutive DNA unrelated to semiconductor gives better results. For the benefit of the continuous attempts at gene therapy of the progress in the area of expanded bone marrow transplants to restore the functions of blood cells when infected with a genetic disorder, and used most of these attempts retrograde viral vectors.

New ways to design a treatment using cells of the organism 104

- Cell culture installed: is converted to the infected cells in organs or in the culture the appropriate gene, and then build a specific structure and in the infected tissues. Thus, cells expressing the gene and provide the onboard product withdrawal.

Diseases which are currently subjected to genetic manipulation

- Cancer: The application of gene therapy in cancer is not now aims to correct genetic defects in somatic cells, but sought to allow the introduction of genes to eliminate them.

- Neurological disease: This treatment is used to reduce nerve damage that accompanies Parkinson's and Alzheimer's disease and to enable the infected neurons to recover their function.

- Acquired Immune Deficiency Syndrome: There are many techniques under the anti- vaccination testing, such as installation and self-inoculation cultures for the primary fiber cells that contain transience's carry code of viral proteins and immunization procedure by genes which are to block viral reproduction.

- Gene therapy of developed AIDS, the first attempt of gene therapy in humans (1990), while the enzyme that remove amin of adenosine make the child is not capable of performing the immune response to resist infections after isolation of lymphocytes from patients, and then insert a normal gent for an enzyme that remove the amine through virus vector.

The treatment of patients through the provision of the necessary genes are still the idea compelling. For those who remain in front of researchers in basic science and conscience, much to be done before gene therapy to succeed.

Fingerprint

That science is progressing dramatically in the current year, so that it can be made in the last quarter of the last century, equivalent to human progress in its long history as a whole. In the field of genetics offers this impressive progress of science and builds on the many hopes in the future of human. While the human in a state of surprise and astonishment, which inherited the technology to adapt the results of gene. The scientists that discover some of the problems to show us the genes which was later named the fingerprint genetic fingerprinting. What are these? What are the issues that can be resolved, and is unable to traditional means of forensic medicine to find a solution? Genes that carry genetic message from one generation to another and guide the activities of each cell is a giant molecule may be converted to resemble the strings, called DNA that contains all the genetic traits, from eye color to the smallest structures in the body. They are the result of genetics in human cells, for 23 pairs of chromosomes in the nucleus of the cell, the chromosomes constitute from DNA. Proteins play an important role in maintaining the structure of the genetic material that lead to reveal the full individual. The finger printing began until 1984, when the geneticist deploy at the University of Leicester in London search of genetic material that may be repeated several times and re-sequences itself incomprehensible represented in the length and location, this research reach in one year that this sequences characteristic of each individual and can not be similar between the two except in cases of identical twins only, and the potential similarity of fingerprints between one person and another one trillion, making it impossible similarity, that was found that these differences are unique to each person just like fingerprints and therefore called the fingerprint genes. Dr. Alec has recorded his discovery in 1985 and named them the name of the sequences of the human person as defined and as a means of identifying a person through the passages approach sometimes called DNA fingerprint "DNA Typing". The genetic fingerprint known through the courts, although had spent time in the detection through forensic medicine, where possible knowledge of this fingerprint to identify the mutilated bodies and tracking children and missing soldiers, as it can mark genes to identify the person until the bulbs of the hair that has been cleared of many of the defendants by identifying the genetic fingerprint of murder, rape and revealed the true perpetrator of the crime, had a genetic fingerprint of the word on the issue of genealogy polarities of a number of issues to prove paternity, rape, and calculates the ratio of the distinction between individuals using fingerprint genes found that this ratio up to about 1: 300 million people, there is one person with the same genetic fingerprint was also found that fingerprint genes inherited according to Mendel's laws of genetics. It has also found that the fingerprint genes vary according to geographical patterns of the genes in the peoples of the world, for example, is different from Asian (Mongolian or yellow race) for the Africans. For the identification of genetic fingerprint requires a small sample of tissue that can be drawn DNA including, for example:

- A sample of blood in the case to prove filiation.

- A sample of sperm in the case of rape.

- A piece of skin under the nails or hair roots of the body in case of death after resisting the aggressor.

- Blood or semen frozen or dry is on the crime scene.

- A sample of saliva.

Ethical implications

That this issue is distinct attention to enrich the scientific research related to this section of the scientific specialization, which is still 106 growing and evolving as a number of questions that put precision together constitute the social issues and scientific issues that require a unique answer response send in self-certainty and a sense of security and assurance. The religions have confirmed the ethics of researcher and research ethics and both sides of the same coin, a search should be moving to the reconstruction and development and preservation of the environment that God created it so well, the search if deviated from their destination and good career development research is not useful and must be liberated with the production and consumption together.

The importance of this subject is first that it does not affect the religious, but very cautiously and in accordance with insights and analysis are limited, and the cure of genetic testing and abortion, infertility and human eugenics and other topics related to the needs of the Muslim scholars to discuss and study and comparison with the fundamentals of the faith and purposes of the law. If we were not the courage and wisdom to show the religious scientific opinion on these issues inherent in our daily lives, will remain controversial among the various currents and contradictory beliefs, which reflects negatively on our future generations and directly affects our faith, one way or another.

Ethical considerations

The gene therapy in somatic cells aims to treat serious diseases, and the possibility of morally acceptable. The gene therapy in cells construction remains a subject of controversy with regard to cell construction cells and with regard there are a number of questions.

- Do we have the right to change the genome of an unborn child?
- Who has the right to approve?
- Are we encouraging the introduction of genes (such as growth hormone) to improve the quality of embryos? Any non-therapeutic uses.

Despite the many considerations of discussions on gene therapy technology, millions of people with one of the different types of genetic disease, they hope to apply these technological developments soon in the attempts to mankind and in the absence of other types of treatment, then should allow the growth and development of gene therapy in somatic cells, under the supervision of bio-security. Which include preventive measures that should be adopted to reduce the need for gene therapy in Muslim societies:

- Interests in genetic counseling in public and private hospitals to help people to absorb health education on genetic diseases and to take the necessary measures. Promoting genetic studies (epidemiological) in families and tribes that carry infected gene and this makes it easier for genetic counseling, as well as gene therapy.
- Do not marry relatives, particularly when it is in the family ancestors are infected with diseases and hereditary.

Medical consideration

It is not preferred to attempts to human gene therapy in the absence of a broad scientific background able to understand the nature of genetic disease and molecular consequences. On the other hand it must be used human gene therapy techniques in the framework of a particular lead to unwanted hard impact and to restore normal cellular function in a person's life that continue throughout the future. It also must be gene expression regulated outside the original a manner as to improve the patient's condition without damaging the cells or the person the future.

The use of viral vectors in human gene therapy is critical concern due to the ability of these vectors to the initiate the particles conditions infected virus that may spread to neighboring cells, or to others in the community.

The treatment of structural cell may cause damage that occur in future generations, and may lead to correct the composition of the affected gene mutations. The remaining operations targeting stay primitive and with non-controlled roads.

There are potential dangers from the use of gene transfer by retrogressive virus but it did not cause any minor damage in humans. The National Institutes of Health has described in the United States malignant T cells in monkeys, but discovered later that these resulted from contamination of the carrier virus.

Religious considerations

- God created man in the best stature and with generosity to other creatures, and tampering with components of the human being and subjected to tests of genetic engineering without a goal is incompatible with the dignity that God bestowed on humans will read on him, "We have honored the sons of Adam".

- Islam is a religion of science and knowledge as stated in the verse, "Are those equal who know and those who do not know", which is not forbidden to the human mind in the field of scientific research and useful genetic engineering in its various aspects in addition to knowledge.

- Everyone has the right to respect dignity and rights whatever genetic characteristics.

- It not conduct any research or carry out any treatment or diagnosis of the genome, of any person unless conducting rigorous and prior assessment of dangers and potential benefits associated with these activities with a commitment to the provisions of law and ethics of this matter and, if not beneficial to health and direct benefit to him. It should respect the right of every person to decide whether he wants or does not want to take note of the results of any examination or genetic consequences.

- All genetic diagnoses, preservation or preparation for the purposes of scientific research or for any genetic examination or its consequences are confidential.

- It is not permissible to offer any person for any form of discrimination based on genetic characteristics, which shall be liable or result of reducing the fundamental rights and freedoms and violating the dignity.

- No research on the human genome or any of this research, particularly in the fields of biology, genetics and medicine, should prevail over the observance of human rights and fundamental freedoms and human dignity of any individual or group of individuals.

- Publication of books should be to simplify scientific information about genetics and genetic engineering to raise awareness and strengthening on the subject.

- The introduction of genetic engineering into the curriculum at different stages and in local media.

-

Futures of Genetic Engineering and Biotechnology and the Internet

Many believe that one of the most important early developments that will emerge from genetic engineering is the technology of artificial viruses, that could become almost as today's design and manufacture of cells and viruses that have changed and stop specific biological processes. Thereby eliminating a particular disease or changes in the characteristics of an individual. Biologists and researchers hope to be making a virus able to recognize the cancer cells and access to arrest its proliferation as well as many other applications and thus allow the virus to replace the surgeon industrial tools and unequivocal chemotherapy drugs, the strongest and the most disturbing and less harmful in terms of side effects

On the basis of current developments in science and technology which they can draw a picture of a bright future for humanity that can be created by itself if it wanted worked according to ethical, social- science concepts. Then will come a duty on humanity as a whole to interconnect network giant relies on a large group of small satellites for communications and electricity will be available in areas. Thanks to remote farms with genetically engineered to convert sunlight into carbon and then to the raw stream and can then run all the equipment and facilities, including communication devices via satellite and the Internet. It had been predicted by many scholars the most important scientific developments of civilization that have been achieved, including information technology and the Internet, artificial intelligence, and the travel space with full visualization and beautiful and optimistic to the future of woven colored threads of science, ethics and technology, philosophy and focus on the short term more than the remote for the application of the Scriptures space because of the failures observed in the draft and the space agencies of U.S. and Russian (Mir station in particular) as well as matter in relation to artificial intelligence.

The biggest and most important feature in information technology and new communication to overcome language barriers and ignored the local culture and traditions and there is no technical obstacles to prevent them from connecting the world and its peoples to each other. But the potential for the delivery of information today has become much easier than the capacity of countries to deliver water and electricity and the provision of medical and housing for their people. Of course, the Internet can not solve all the problems of the world's social and economic development, but we started to see positive effects in many areas and a variety of no conceivable when one plans for the network. It is a positive revolution which imposed itself.

To illustrate this category which used to computers and networks of those who daily through the Internet and its information online and link part of their careers and those already receiving significant superiority to the other categories and who update the world via the Internet are in transition to a new server-class gap between the connectors are growing rapidly.

Either with regard to biotechnology and genetic engineering then it is of course quite different, the world has witnessed rapid progress and as a surprise to us and put the sheep reproduce and international human genome project at the forefront of current scientific. In our vision will bring us a greater developments and surprises in our lives from the Internet and genetic engineering, especially after the human genome and not from the sun or space.

Two recent examples will come believes to many of the biggest scientific surprises that have occurred over the last few years, first cloned sheep and the international human genome in front of the computer any more than artificial intelligence to human intelligence for the first time. Some people view the prospect of human cloning possibilities and tremendous results and violating some positive, some seriously, in the social and cultural heritage will be able to parents as soon say- the possibility of using cloning technology and genetic engineering to compensate for specific genes to their children before their training. This will change the capacity of children's physical and mental to protect from certain diseases and symptoms and arming the other capabilities to facilitate their life around them. But this technology in the first decades at least will be expensive and most likely this will lead to expansion of the difference between the two layers of rich human or genetically grafted and natural. There is no doubt that this will push the old division of mankind into masters and slaves, unless you turned this technology accessible to everyone, and this is expected.

Philosophy

It believes that the term philosophy was a Greek origin, and consists of two words (Philo-Velia) and meaning of loving and (Sophie-Sofia) their meaning wisdom, either in the Arab-Muslim heritage has been found to have several expressions, including ethics. Philosophy of the most important specifications that fields wide and several kinds of sequential steps leading towards the formation of thought essential to simplify the perceptions and understanding of human dilemmas. But they are all looking for the nature of things by using the mind. It has spread in Greece a lot of contradictory intellectual movements philosophy of idealism and realism The former see the values of fixed and virtues do not change, while the second sees the care of the senses is more important than a focus on fantasy and conform to the ideal in the fact that the virtues fixed, educational philosophy is required for the fact that education is part of the fully human existence Calvin, science and language.

The concept of education is different from, for example, according to the philosophy that deals with the type, indicates that education is the formulation of the same human being announced and goodness is believed Plato that education is consistency between the soul and body, while Al-Ghazali believes the education her priority is the spiritual and humanitarian atmosphere. And lead exemplary education to the highest degree of maturity of the children, either natural philosophy refers to the mental preparation of the child. While education when existentialism that man is free and is subject to the inevitable and philosophy of pragmatism as vision Dewey suggest that continuous education organization of experience and adapt to the social reality. And John Dewey, one of the philosophers who pointed out the principles on which the modern concept of education and the education of them small community which comes to life and continuing education forever and curriculum must keep pace with the life and mission of education is to prepare the individual for life.

Among the Arab philosophers who were interested in the problems of education such as Ibn Sina and Al-Ghazali and Ibn Khaldun. Son Khaldun (1332-1406m) confined to educational principles with him must be gradual and the transition from the known to the unknown and from easy to difficult. It particles to colleges. The European Education, which Pflasvetha including Jean-Jacques Rousseau (1712-1778m), who called for equality and return to a normal life away from the corruption and meet the needs of the child. And Herbert Spencer (1820-1903m), which focused on the psychological concept of education and private psychology.

The subject of philosophy is knowledge () and knowledge of the natural facts (relative) standard and the facts (values and ethics) and Tat

Idealist philosophy Idealism

And it linked to this philosophy pioneered by Plato, which is the perception of the existence of two worlds, the world of ideals (hard) and the real world (variable). The community consists of two layers, one thinking and other works In other words, the first class is linked to the educational framework for the purpose of access to knowledge and that requires the mind, and thus Knowledge is that link the mind to be real and immutable. This philosophy and believe in basic principles of strong belief in the existence of an independent in a perfect world the real absolute ideas.

This philosophical school follow a curriculum steady N to develop. The idealism and teaching methods are based on the basis of mental training staffs. According to Applied features of this philosophy in the field of education, the accumulation of knowledge approach is clear and unchanged and adoption of tools learning as ways constant teaching and exams without the use of traditional means and focus on the mind and the lack of school trips to adopt and approve of corporal punishment and finally that learning is the focus of the education process, as well as it sees Plato and teacher Socrates (470 BC -399 BC) to fixed values and virtues do not change and that education must take care of reflection and imagination and that human nature is composed of bilateral spirit and body and must take account of these bilateral upbringing of the individual in society as righteous. According to this unrealistic perceptions of Education is must transcend the spirit without neglecting the body. And it provided the ideal philosophy for education and training of the Socratic idea generation which is based on stirring the mind and push for self-search, add to that Vehtm education when idealistic philosophical issues experimental spirit with abstract thinking and save the information practiced by high school.

Realism philosophy

According to this philosophy of Braidha Aristotle (384 BC -322 BC) that there is only one world is the world of reality is characterized by fixed principles and senses care is more important than Turkao imagination and consistent with the idealism in the fact that constant virtues.

Follow the realists narrators approach stability depends on continuing this philosophy to discover the universe and the world and work to understand the laws in force, which includes all of the facts within a stable and consistent world attempts. Thus Valoajpat placed on realism philosophy for Educational clear according to the following:

•an approach which accumulate natural, social and cultural knowledge.

•acquire the knowledge, skills and habits to prepare students for life.

•Prepare specifications teacher distinct process.

•extracurricular activities necessary realism of the school.

•learner is the focus of the educational process.

•extracurricular activities is an important part of the curriculum.

•curriculum of a set of facts that scientists have detected consists.

•The use of programmed instruction machines.

And the fact that the realist school Qdj brought down the senses and meted out the philosophy and ideals of meditation and fantasy to reality and the senses so that scientific knowledge and curricula astronomical and mathematical Kalaom Square widened. And John Locke (1632-1704m), one of the pioneers of realism philosophy and endorsed it by his conviction gave critical thought ample room and the experience and the reality and the senses no example and abstraction based on knowledge and science he sees that the child a blank sheet draw Storha of fact, as well as it has added Luc need to study natural phenomena, along with other math and science.

The Komenus (1592-1670m) realist sensory Making of the image of the most important methods of child education in schools, along with facts enthusiasm and believed to be the founder of the first special education child of freedom. And the development of sensory approach to teaching and took care of the physical and moral education to both.

Finally, the realism reached a multiple convictions, including:

•established curriculum experimentation and exchange of scientific and systematic doubt (the road to see the existence of God, in real life.(

•encouraged the learner to observe natural phenomena in an orderly fashion.

•called for acts of the mind in the analysis and the independence of the senses.

•This school did not distinguish between the world of ideals and spirit, between the world of seriously indeed.

•did not elaborate on the mental meditation.

•focused on professional education.

•called for linking the curriculum Balhajiat life.

Pragmatic philosophy

The roots of this philosophy go back to ancient times and the writings of Hrakulais (535-475 BC) and Undtlaan (30-95m) and Harzubayrs and William James (1842-1910). While contemporary pragmatism linked to the New World (the second half of the nineteenth century and early twentieth century) called pragmatism generally Baladhatih and pragmatism, development and operation.

Of the principles of this philosophy that education is life and tied the education community service philosophy giving Alentalm deal of freedom. And the human organism adapts to the environment and that the biological world is relatively constant in flux and the truth is absolute and variable society and democratic decision-making. John Dewey has established the features of this philosophy.

Pragmatic philosophy role in the development of teaching methods and by improving the traditional way and in a manner trial and error, or experimental way. John Dewey School-based Albergmteh philosophy the intellectual revolution against the historical and traditional schools, which focused on the information and believes firmly values but relative morality is renewed according to the convictions of the community.

The teacher in the philosophy of pragmatism he does not teach traditional materials in a systematic way and the teacher moves from idea to another in a sequential manner and deal with all the idea on the grounds that it in itself and suggests future problems for his students. And that the values vary depending on what they put the time of the convictions of the community and the individual.

Education at the pragmatic philosophy is not broadcast knowledge to the student for the process of knowledge, but to help him cope with the social environment and the needs of the student was able to stir up the forces as required by social attitudes. And rely perceptions pragmatic philosophy on building curriculum as confirmed by the student on what he wants from him, and using student Alqrah, writing and arithmetic as a means rather than targets, and the curriculum is interested in the facts relating to the nature of the child, also confirms

Islamic philosophy

There are major Islamic philosophy perceptions reflected on education, and these perceptions that the Quranic verses explain the nature of the universe and man, knowledge and values, and everything in the creature universe to God and the changes that occur in the universe is governed by paths and rights in Islamic philosophy has marked special was the gift of the Creator stature over the place that was given and all the other creatures in the universe to create the service.

Education is happiness in this philosophy and that the child has whiter page where there is no hindered Education In practice Education is beginning the science of the Koran and the chatter beyond are going to learn science has refuted Ghazali arguments philosophers in his book, The Incoherence of the Philosophers, while triumphed Ibn Rushd philosophy wrote The Incoherence of the Incoherence Canadian and longer philosophers who emerged in Islamic history.

Natural philosophy

This philosophy believes that man is good by nature and what is human spoil the society and its institutions and education have a target to allow for the natural growth of the child, as the jam Ptsourath negative and positive learner childlike nature. This philosophy also believes that the senses sources of education and outlets for the development of thought and not a role for cognitive balance.

Natural philosophy appeared to calls by reference to the educational activities that are aligned with the natural laws of indigenous cared Rousseau (1712-1778) these ideas and crystallized natural philosophy and Allamadrsah cemented the role of nature in the development of children in terms of:

•freed the child from school and classroom activities.

•meet the needs of the child and the need to remove obstacles facing it, according to environmental requirements.

Rousseau was one of the first to call for self-learning and the nature of the most important educational principles that are consistent with, and that women found the man to please, and this philosophy advocated the need to respect the individual and the protection of children from societal pressure differences.

Existentialism

Kdeckart existentialists and Sartre believed seriousness and absolute freedom, humanity, a philosophical vision of human existence and the responsibility and the nature of humanity and the world, knowledge and values existential emerged in Europe after the First World War (1914-1918) in Germany and then in France. It is believed the existential education as perceived by them to human existential indoctrination and does not accept the promotion of education for the existence of man and literature, music, philosophy and the arts of the general requirements, dialogue and debate and the basics of teaching methods and programs stable unacceptable.

Philosophy of Deschooling

Each of John Holt (1992) and appeared to Cash (1994), the pioneers of this philosophy, which depends entirely on educational institutions Some believe the philosophical basis for this school is Gandhi's theory of education, especially in environmental education and the foundations of this school:

•learner be close to its environment and interact with it.

•learner learns from his peers.

•Achieve learner itself.

It is believed the others of the pioneers of this school that limiting education within the school not feasible and therefore extracurricular activities requires an important role in our schools and to be environmental education general philosophy across the entire curriculum, not academic subject separate philosophy Allamadrsah is the return to the simplicity and comprehensiveness and return to the educational roots .

The philosophical concept of the mystery of life
There are three philosophical concepts of the world and developed as a result of human intellectual effort spiritual concept and the concept realistic materialist conception realistic divine. It can be evaluated and try to rush to one of them, or the formulation of the concept of the center between them. The conflict between the divine and the physical manifestation of the conflict

between idealism and realism and that the concept is a philosophical world one of two things ideal concept materialist conception does not correspond to reality at all, realism is not according to the materialist conception as the ideal is not the only thing that is opposed to the materialist conception, but there is another concept of realism realistic is the concept of divine. The concept of the divine to the world does not mean cutting out natural causes, or to rebel against something Facts sound science but it is a concept which is considered the deepest cause of God. The material back in the door is the perfect argument to the field of spiritual and either the spiritual concept of the divine way of looking at it is the reality.

As for the scientific field, there is no God and material Valfelsov whether divinely or materially believes in the positive side of the flag, there is no question in the scientific philosopher, my God another material, but there are two Filsvetan and in conflict when it was a matter of being beyond nature.

Valalhei thought to the fact that just about art, any outside experience and physical deny it is believed to be natural causes revealed by experience and extended her hand of science is the primary reason for existence. The Divine is believed that the human soul and the (I) with abstract art and perceptive and thought phenomena independent of nature and art.

The nature of the directory you can divine that is provided by the mind, not of direct experience unlike the material traditionally regarded as evidence of experience on the concept, we believe that the concept of divine or metaphysical issues in general can not be substantiated experimentally. Materialism need proof on the negative side, which distinguishes itself from the divine and it's the direction of philosophical Kalalheih, because science alone does not prove that the material of the concept for the world to be a material process but everything undisclosed knowledge of the facts and secrets in the world of nature leaves room for assuming a higher reason above article. If we look to a set of basic concepts about life and the way of thinking which can then be first addressed to the theory of knowledge and secondly philosophic concept of life and when studying the theory of knowledge can focus on relying on the mental way of thinking which represents the necessary knowledge on the experience as well as the study of human knowledge value on the basis of logic mental not physical.

Epistemology mystery of life represents the perception and the ratification of which reflect the cognitive The former represents a presence, such as heat, light and sound, while the second (ratification), which represents the recognition that the heat energy source, for example, the sun and other concepts. A number of theories have been dealing with these perceptions, including:

•recall theory

The (previous recall of information and that the human soul exists independently of the body). They isolation from the article, and can correct some of the mistakes of this theory, which represents that the soul is not something that exists in the abstract by the presence of the body but of the core material movement.

•Mental theory.

Put this view before (Descartes) and (it was) and indicated the presence of exporters perceptions firstly sense (heat, light, taste, etc.) and the second instinct (the human mind has the sheen and perceptions did not emanate from common sense, but is fixed at the heart of instinct). Vlachtlavat with the recollection theory is the fact that the source of the sense of understanding of the concepts, not the only reason. And the drawbacks of this theory is to return the entire cognitive sense.

•sensory theory

Based on the theory of sensory experience and common sense in this theory is the infrastructure, which is based on the base of the human perception.

The Futurism and Science

Studies that rely on scientific prediction

These future studies focus on several areas of operation physics, life sciences, mathematics and sparked actors on the scientific and industrial institutions and other such studies and indicate to the rule-based industries of science within the terms of reference of renewable energy and genetic engineering techniques and life industries and electronic industry Computing and transportation and communications, space and materials science.

Researchers from these studies are expected, for example, the use of satellites for the transfer of energy by transferring solar energy to microwave wavelengths can be transmitted to ground stations and then converted back into energy that can be utilized. But in the field of genetic engineering and remember a lot of scientists and future perceptions of the following in this area:

•clone of genius or free of disease rather than relying on coincidence the advent of births may not be a genius.

•production plant produces roots potato tubers, while the plant itself produces Tomata.

•In the realm of electrons in the computer industry:

•The computers in the future to help the doctor make the necessary his examinations such as blood analysis and other tests and then diagnose the disease and provide necessary medication to patients.

•The computer in the entire private houses manage those inhabited by people with disabilities home and turn on the TV and cooking and catering by robots.

In the transportation and Telecom rights in the future will be able to do with:

•lapses without going to the university, but in front of his computer.

•scientific conference attendees to thousands of kilometers away, and participate in the discussion without having a physical presence in the conference.

But space scientists doing great achievements on the level of outer space, including:

•the establishment of satellite settlements on Mars and the moon and the atmosphere that surrounds the ground.

•set up factories satellite numerous electronic components and medicines produced (outer space is very useful for the pharmaceutical industry is pure.(

•freezing of embryos and placed on a spacecraft to form generation and directed embryos from freezing conditions to form the second generation, and so man continues to send his grandchildren to the ends of the universe.

Either in the materials science of man will be able in the future of the industry materials with distinctive characteristics can be used in clothing and the automotive industry as well as the use of carbon instead of silicon in the computer industry of life instead of silicon computer.

In a study on the future of the world, after four hundred years of perceptions it can refer to them for the future, namely:

•severe pessimism.

•pessimistic warned.

•optimistic warned.

•enthusiastic about the growth and technology.

Severe pessimism

This perception is supposed

• The technology would have little effect.

• that man does a great depletion of natural resources.

• income disparity between the developing and poor countries will increase.

• Food decreasing.

Pessimist warned

• growing population will lead to diminishing returns.

• depletion of resources.

• Increased pollution.

Optimistic warned

For this model

• Resources growing as a result of technical progress.

• rising standard of living.

• sanity in the consumption of resources and the preservation of important resources.

• Future Shock

It will create in the future of new generations infected with the so-called future shock according to the following scenarios:

• Future Shock acute illness experienced by increasing numbers of people (disease inability to adapt to rapid change(.

• responses for future action based on what you know Daalh in the ability to adapt.

•multiplication of psychiatric and neurological diseases with a lot of people.

•The technological progress has been creating facts on our awareness could not be absorbed.

•human future will be expatriate man without roots and without confidence amid the dunes of quicksand.

You need to study models for future many things, including:

•database includes comprehensive statistics.

•a set of preliminary studies that rely population or resources.

•adoption of specific scenarios and targets serving scheme.

•Computers with a great capacity for storage and handling information.

Most of the current and future studies completed in 2005, as required for further studies:

•technical means such as computer to store and recall the many information.

•Use of special programs for the forecasting process.

•Many experts, technicians and programmers are not from one's jurisdiction.

•adoption of the current concepts of the Industrial Revolution (second) that emerge from science who founded the various discoveries.

−Chemistry contributed and developed mainly to new technology production processes.

-Mathematics and physics at the foundation of nuclear fission and the invention of the electronic calculator.

-Life sciences effectively have an impact in the fields of agriculture and medicine.

-Associated electronics industry computer industry.

And we're seeing scientific advances in chemistry and life sciences, mathematics, computer sciences and other sciences is increasing every year and became the basis for most of the other sciences. It is possible that future studies in science is divided into three types:

-Studies that rely on prediction.

-Future studies that rely on intuition.

-Future studies that rely on statistical information detailed within the sports programs or computer models.

In light of this, there are four scenarios can refer to them for the future, namely:

-Severe pessimism.

-Pessimistic cautious.

-Cautiously optimistic.

-Enthusiastic about the growth and technology.

The human being was thinking of his future, his primitive tools that were addressed neither argue his aspiration nor prevent him from seeking the future, then these tools were developed and become available and necessary to study and predict the futurism. Later, we noticed the scientific advancement was growing every year in Chemistry, Biology, Mathematics, Computers, Medical Sciences and Engineering Science.

Huge developments have taken place in Chemistry and Life science flourished new branches in Chemistry and witnessed tremendous development in the study of Molecular and Atomic structure, such as the use of Lasers and low & high energy X-ray. The Chemistry of life has merged, including molecular Biology and chemical Biology that have a great future, also got tremendous developments in enzymes either in the life science and includes dozens of science specialists subsidiary, has used genetic engineering applications in areas of medical, agricultural, industrial and vaccine production and hormone treatment of incurable diseases.

Human being was thinking of his future, his primitive tools that were addressed neither argued his aspiration nor prevented him from seeking the future. These tools were developed and became available and necessary to study and predict futurism. Later, we noticed that the scientific progress - in Chemistry, Biology, Mathematics, Computers, Medical Sciences and Engineering Science - was growing every year.

Huge developments have taken place in Chemistry, and Life science flourished new branches and witnessed tremendous development in the study of Molecular and Atomic structure, such as: the use of Lasers and low & high energy X-ray. The Chemistry of life has merged, including molecular Biology and chemical Biology that have a great future, also got tremendous developments in enzymes either in the life science and includes dozens of science specialists subsidiary, has used genetic engineering applications in areas of medical, agricultural, industrial and vaccine production and hormone treatment of incurable diseases.

In the future, the peoples will be able to participate in the materials industry to produce materials with distinct qualities that can be used in the industry of cars, garment, as well as the use of carbon instead of silicon in the manufacture of bio-computers, and in other areas there will be success in the war against disease. Scientists expect that they will know the secret of the cancer, then to destroy it; also to understand old age, and then to find the means to prolong life.

In about two hundred years, there will be four possible scenarios for future reference, namely they are:

- Very pessimistic.
- Cautiously pessimistic.
- Cautiously optimistic.
- Eager for growth and technology.

In future new generations will be created and involved, the so-called future shock and Toffler goes through analysis that includes:-

- The future shock is a severe illness that affected increasing numbers of human beings and can be called satisfactory inability to adopt to rapid change.
- Reactions to the future depends on what is known about the ability to adapt recalled hypothesis in the 1st. instance, that the proliferation of psychiatric and neurological a lot of people due to the shock of the future and technological progress and scientific facts have not been able to create awareness on assimilations.

Scientific Incubators

The world economy also reminds us, that it is moving towards a knowledge based economy and this trend is explained by economic theories, including new growth theory. This theory says that sustained growth (rather than growth for a short period), is directly dependent on three factors:

- Technological level.
- Technological growth rate.
- Saving ratio.

The traditional factors of growth, in which capital and labour are involved in directly in the growth equation. These changes - which are very important in the world economy - are considered as added value that comes from the high technological level, the technological growth of the state and not only from capital investment and work forces.

Economic growth depends certainly on the researchers, discoveries and inventions and to those who invest these inventions and creations. This is in addition to the good implementation of these creations that depends on skilled workers, who can deal with modern means of production.

Many countries are no longer depending solely on industrial zones and free zones, because this mechanism does no longer believe in acceptance of added value. These countries began to adopt new patterns since eighties, such as, technological areas, science and technological cities and areas of knowledge. The objective basis for these patterns is the maximum utilization of new and innovative ideas which emanate from research centers and universities. The link between research and development on one hand and industrial and service activities on the other hand, is the nucleus of these areas or new cities.

The technological incubators are those of new patterns that were adopted for the purpose of achieving the objective mentioned above. Technological innovators are effectively helping to move the idea of a new form of laboratory or experimental or academic to the model proposed for breeding technique. The incubator creates the company and owns shares of the company's budget; the technological incubators are the best means of developing countries or developed countries to encourage the establishments of the initiators of their companies and are product of proven economic.

The technological incubators are of the entrances adopted at the global level, to encourage and support small and medium industries, where there are today more than 1500 incubators operating in the world focused in the USA, Europe, Japan, and there are 500 incubators in developing countries. There are about 200 incubators in France and more than 100 in Britain and 200 in Germany. Furthermore there is Japanese experiment in the field of incubators and science parks.

The communities in general must use the modern technological developments and the appropriate environment form. These communities, which would deepen the work of free thoughts, contributing to industry change. Among these methods are business incubators and small projects. The business incubators is the integrated system of small projects that are, at the starting stage, in need of special support and protection so as to enable them later to move to foreign labour markets.

The overall objectives of the incubators are the development and creation of innovative projects, assisting owners of innovations and inventions, providing support and funding and providing services to destination funding for research and knowledge. One of the international experiences of such incubators is to implement the ambitious strategy for the developments of small projects; it requires a central body to manage, implement and follow up these strategies. The American experience is tracking U.S presidential administration but in France and Malaysia through specified ministry. In Egypt the incubators is represented at the

social fund for development of the presidency of the council ministers.

There are currently in USA over six hundred of technological incubators such as Austin technological incubators, were used to reduce the failure rate for new projects, 50 projects have been graduated from the incubators, and 10900 new jobs were developed.

There are many types of incubators according to the objective for which they were established, including:

- Regional incubator
- International incubator
- Industrial incubator
- Specific incubator
- Technical incubator
- Research incubator
- Virtual incubator
- Internet incubator

The requirements for admission to incubators are place for the project, financial support, and technical support and skills developments. The incubator is preferably be located adjacent and not inside the campus of a university or research center in order to benefit from the resources and applied researches, laboratories and workshops, services and professors.

.

Chapter Three

Techniques used in science

It is used in the separation and diagnosis of many chemical compounds, biological, etc. There are many types of chromatography including adsorption and ion exchange retail and gel filtration and other technical methods for the purpose of use paper and thin layer and gas chromatography.

Chromatography include multiple ways that all based on the separation of compounds based on the difference in migration through the passage in the center of force, as well as the tendency to face hard "Stationary phase" for central transgeneration and face the hard nature solid or gaseous, or liquid depends on the tendency of various materials to hard to face multiple methods such as adsorption "Adsorption" ion exchange "Ion exchange" may include all kinds of chromatography of these methods.

Based on the tendency of the ions or molecules to materials other than for mobile and non- soluble, which owns the distinct shipments, or molecules that carry one or more of the positive charge exchange with the positive charge associated with the Ionia face, the mobile Resins "Resins" with a negative charge is called this ion exchange process with the positive charge "Cation exchange" and reverse ion exchange is called a negative charge. Examples of the reciprocals of the ion non- animated "Immobile Ion Exchange" that are used in chemical research.

Polystyrene where will attend the multi-way styrene polymerization contributory "Copolymerization" with composite "Divinyl benzene", which adds to the styrene chains cross- shaped multi- written and added then aggregates the active ions altering the chemical composition of the original units that can be prepared for example, styrene resin that contains strong acid groups such as SO_3H hold process of "Styrene-Divinyl Benzene sulfonation". In the same way can be prepared that contains the totals as strong as NR^{3+}, or weak acid groups such as the COOH groups or grass- roots groups such as NH^{3+} and types of preparation depends on the ion concentration reciprocals composite "Divinyl benzene" the amount of strings cross referred to the number listed after the name of the resin, such as "8X" Dowex 50, which contains 8% of the "Divinyl benzene".

The type of chromatography by gel filtration on the difference in the movement of compounds during the gels with regular pores partially used for the purpose of the way in a column filled with from one type of granulated gel filtration.

The pores is able to expel particles with, partial weights more than 10000 if we, for example at the top of the column a small scale solution of dissolved protein and molecular weight of 70,000 with ammonium sulfate the following happens:

- Protein molecules expelled from the pores of granulated gel filtration.
- The migration of protein size start "void volum" outside the granules column.
- Interference ions NH^{+4} and SO^{-4} small pores of the gel granules nomination so there is the amount of liquid required for the expulsion of these molecules outside the pores.

The granules are gel filtration where proteins are separated by major united ammonium by successive periods of time and be dependent on the size of the separation- free liquid, gel filtration is very important ways often used to separate proteins from salts "Desalting". Electrophoresis is applied techniques migration electric technology application. It is accurate in the middle of an insulator as a result the electric field moves in a minute steady pace, and can then measured the balance between the electric power Eq, and the light of that we get the equation.

The gel gromotography is the primary means of separation and purification of various enzymes and proteins, as well as the fragmentation of nucleic acids and proteins in the treatment, especially when quantitative diagnosis of some human diseases, as well as by the method of the exchange of tritium to test the protein structure or the structure of DNA and a study of the link between proteins and small molecules. The thin layer chromatography is mainly used for amino acids, polysaccharides and simple sugars, fat and various steroids and other small molecules. The ion exchange chromatography applications, including chromatography cellulose DNA, to purify proteins associated with DNA to separate in general the Bio-compounds according to molecular weights. The affinity chromatography is used to purify the enzymes and antibodies and transport proteins, membrane proteins and the chips and sugary proteins and the separation of animal cells in particular.

Immunological methods

The developed immunologic tests, which are used to estimate small quantities of antigenic non- radioactive compounds in a mixture of large numbers and quantities of miscellaneous materials testing immune- ray equivalent to or greater than the sensitivity tests of natural chromatography. That the purpose of radio-immunoassay test is to assess the basic particles:

- Those that were not labeled by radioactivity within the body by proper specific activity or without adequate labeling of other compounds.
- Those do not know the identity, it could interact jointly and thus to compete with the antigen known.

There are many substances that are measured in the test radioimmunoassay, including the hormones and pharmaceutical agents, vitamins, and factors assigned and materials in the blood and viral antigens (viral) nucleic acids and nucleotides. In some cases, there is no acceptable method for the preparation of labeled antigen, which is permitted by measuring these materials using immuno-radio metric method, which can be measured by an unknown amount of antigen directly, through its combination with specie labeled. Examples of immunological methods that used in the biological tests of:

- Diagnosis and weakening the various types of bacteria by agglutination.
- Diagnosis of (viruses) by inhibition of virus generated by agglutination of red blood cells especially the antibody present in the serum.
- Diagnosis of gonadotrophin in the urine of pregnant women by testing the inhibition test and complements.
- Measurement of making DNA in phage infected by complement fixation.
- Diagnosis of relations between proteins by specific reaction.
- Diagnosis of tumors.
- Testing materials of clinical importance in children.

Immunoassay

The development in many areas of clinical medicine by "Yallow & Berson", which developed a radio- immunoassay technique "RIA" for the purpose of measuring the concentrations of very low- lying materials and the offer of "displacement" antigen, which is marked by radiation from the body own by adding increasing concentrations of antigen, the record is marked by radiation and this applies well in the science of hormones, as the hormone levels in the bloodstream always be very low, making it difficult to measure ways of life and conventional chemical. There are many hormones that can be measured easily and quickly test by radiation immuno assay including, prolactin, which was found to be associated with spinal tumor glandular "anterior pituitary gland tumor" and more permanent with symptoms by menstrual "Menstrual disturbance". In fact, the measurement become a part essential to the tests of modern futility "infertility".

Measurement by radio- immunoassay and other methods of link

There are three methods of test for the purpose of measuring the materials of life:

- Biological assays.
- Binding assays.
- Physical chemical assays.

There are also two types of tests in cases of binding namely:
- Test the link "Ligand".
- Tests of Binder.

The test of linking section are all kinds of bands that can be used, namely:
- Cell receptors.
- Circulating binding protein.
- Antibody.

The types of acquisitions "Tracers" used are included:
- Particle.
- Fluorescent.
- Enzyme.
- Isotope.

Applications of the main principles of radio- immunoassay test

Radio-immunoassay test depends on the competition between the antigen, which is labeled and non-labeled sites on the anti body component complexes ratio on the amount of antigen without radioactivity.

Using these antibodies tagged for the diagnosis and distribution of antigens by the optical microscope or electron microscope in tissue sections and sandwich technique used widely. The unlabeled antibody is 88 placed on the section washed with labeled antibody increases and the layer enzyme linked or marked by "fluorescence" against immune "Immunoglobuin" and can therefore be signaling antigen under examination.

This technique in applied research (Bio and medical research) Examples using antibodies "antisera ordinary polyclonal" on the "Topographical mapping" of the various types of cells in tissues such as "islets of Langerhans".

The immuno tissue chemistry containing various anti blood serum describes beta cells of containing insulin, located in the central mass of the island "islet" cells while the A cell that secrete glucagon in the peripheral side linking them to the cells D secreting "somatostatin". An examples on these the use of monoclonal antibody in the clinical diagnosis of tissue "histopathological" of the disease when the test of the "Lymph nodes biopsy" where it help in the classification of a certain type of lymphoma "lymphoid tumor" (e.g. Hodgkin's disease and various types of lymphoma "Lymphoma").

These antigens are characterized by being glycoprotein's present on the cells, especially white blood cells in humans and is then called human white blood cells, "Human leukocyte antigen" (HLAs). Carrying this antigen genes of the immune response status, where they control antigens were present in different tissues in the body, in addition to genes, there are mismatch humoral immune response "Humoral" and cellular "Cellular".

Immunodeficiency resulting from the lack of a genetic condition in the inability to create an immune cell, or one of its outputs and symptoms of the disease- causing immune deficiency commensurate with the degree of destitution and accompany him.

One of the examples on the case the disease resulting from this deficiency (AIDS), "Acquired immunodeficiency syndrome" (AIDS) or acquired immune deficiency syndrome and was attributable to a virus of the type of regression "retrovirus" with a tendency to lymphatic cells "T" in humans, called "Human T cell lymphocyte".

This virus has several methods to spread such as blood and mucus and the interface is accompanied by injury to the virus to many diseases, so called on the situation of the disease and syndrome of one disease. The assumption is based on the perception of geometric depends on logic that a collapse of the immune system caused by AIDS, produced by

127

relationship engineering between HIV (human immunodeficiency) in the body of the infected and the immune system where it is can do the following:

- Cloning of the virus rapidly that destroy large number of cells of the system as there are two types of tests in cases of a binding.

- Faces a viral reproduction for many years through every defensive response to prevent the virus from reproducing.

- An imbalance in the latter for the benefit of the virus "HIV" event leading to AIDS.

- It can show new geometric forms of the virus as a result of mutations that be able avoid the defense forces of the body in some way, and confuse the immune system, which enable many patients to stay healthy for many years, finally collapse due to the boom continued, speaking of the virus.

Diagnostic Imaging

In medical diagnosis it is adopted mainly on the knowledge of diagnostic imaging technology spectrum, including the use of X-rays and gamma rays from, which is characterized by being electromagnetic radiation ionizing radiation, then began to think about using the term of this non- ionizing radiation infrared or microwave radiation and technical NMR magnet. The examples of spectral techniques used in diagnostic imaging:

- X- rays
- Gamma- ray
- Ultrasound
- Infrared
- Anti electric tissue
- Visual mechanisms

X-ray

The oldest techniques that is used in diagnosis and therefore will not focus on the importance of being where they were getting on the first pictorial representation of various tissues obtain after the development that is built on a limited computer assistance.

Gamma rays

The purpose of gamma-ray is the imaging profile then it was developed as computer- assisted also in the eighties which was called "ECT" and was then developed using imaging "Postiron emission tomography (PET)" where the radiation of tissue is carried out by position (positively charged) and thus can get a picture to clarify the life processes of the tissues that carry electrons and draw.

Ultrasonography

The speed of these waves are characterized by being less of electromagnetic waves, which provides an opportunity to measure the fetus as well as during the stages of development in the womb, added to that the fact that this technique is based on the fact that the X-ray is not ionized therefore it is not a preferred use in diagnostic imaging

Nuclear magnetic resonance imaging

Despite this technology it is old, but it was then developed for the purpose of medical diagnostic imaging has gone from the seventies, where the nuclei of atoms is measured by the disposal of certain substances found in different body tissues. The criterion for the disposal of these seizures depends on the radio pulses that are similar to the frequency in the field of outer-core magnet and thus to obtain a diagnostic can be used.

In the medical applications for the purification of nuclear magnetic resonance imaging to obtain imagery of infarction that occurs in some parts of the brain and important developments in this area the integration

of multiple techniques and access to advanced apparatus for nuclear resonance imaging, including the "TMR" and "MRI".

It is important experiments that experiments are used the magnet resonance imaging of kidney transplantation, which was filmed nearby parts of the kidney and then infected the interactions that take place within the body after transplantation and efficiency of the cultivated parts. As well as imaging of tumors within the liver and liver imaging at the time of myocardial fibrosis or within, as possible, filming parts of the stomach and colon and to identify tumors. It was also to obtain information about stroke and is believed to imagery obtained of cancerous tumors of the brain were more pronounced than the use of X- ray.

It can be measured by any inflation occurs as a result of heart disease, and can also study the problems of the heart due to the presence of any obstruction or infarction in one of the blood vessels and could also portray the evolution of stroke, heart attack and its impact on the heart.

A nuclear magnetic resonance imaging "MRI"

This device is used which was created as a result of the development in the technology of magnetic resonance spectrum by the registration of spectra of life processes taking place within the animal body where the magnet-making with full slot by placing the human within the magnet and thus these devices provide a complete picture of the part which is conceived, and the advantage of the fact that this device magnetic field is not harmful, and the microwave radiation used is not harmful too.

It is possible through this device to study the effects of ongoing parts of the human body while taking a particular medicine can also be follow-up of the various core elements and sequentially, as well as to study the changes occurring stereoisomers of chemicals inside the cell as a result with other molecules.

Labeling with radioactivity

Require a lot of chemical analysis revealed small amounts of material with amount of concentrations $10^{-4} - 10^{-6}$ molari therefore it requires the development of other ways to respond to the concentration of low- lying, such as the development of experimental methods by radioactive to solve many of the other problems that might face them. Some of these methods that could be used by dual- labeling for follow-up of two similar materials formed at various times by pulse method for follow-up fugitive substance at a time after the configuration without interference of other material. An example is the use of radioactive materials in the chemistry of life:

- Choose a material that resides on small concentrations, which are difficult to measure by direct chemical methods.

- Distinguishing similar molecules in different chemical sites.

- Analysis of mixtures that are very complex, which can not be done by various conventional chemical methods. Including:

– Enzyme interactions (DNA polymerase).

– Measurement of molecular weight of the DNA by labeling the final group.

– Diagnosis particle by settling with the anti body. Protein purification, which does not have a chemical test.

– Diagnosis of active centers of enzymes.

Autoradiography

This method is used to detect and locate radioactive materials in the cells or tissue for example, and so the molecule itself and is done by the impact of radiation emanating from radioactive materials or emulsions of photographic plates specially designed for radiation imaging device self-motivate, where silver halides grains, located in the emulsion as a result of the dissolution of radioactive materials in the sample, and the emission of radiation, including activation and work output reduction as indicators minutes for the site radiological effectiveness.

Signaling models resulting from the grains chemically and radiation efficiency in the presence of structures that are in contact with these granules and the microscope can be obtained from the resulting image on the two types of information at the site of radioactive materials and the quantity of a radiation of as the amount of silver particles is directly proportional to the severity of radiation present.

Of the modern applications of this technology as follows:

- Measurement the number of molecules of DNA in bacteria phage.
- Measurement the number of secondary units of the chromosomes.
- Double vision in the DNA molecule of bacteria.

Protein engineering

It is the technique that allows the installation of structural proteins desired in order to build a clone- mediated DNA "Cloned DNA". There is no relationship between the latter and engineering of proteins used, including the building of protein functionally, chemically and physically.

The DNA could be modified by two ways using:

- Mutagenic in private venues.
- Switch sections of the nucleotides.

The protein engineering include modify the structure with protein mediated by genetic engineering and most protein engineering is carried out currently in the field of enzymes, either to speed up its response to the incentive or to become more receptive to acid and heat.

Example: "Cloning" the cDNA for the receptor of "acetyl choline receptor" facilitated the technology which is called site directed mutagensis for getting sequences skilled "Deletions" or substituting some of the amino acids in an additional unit "subunits" of the receptor and then it can test these changes on the functional aspect, and are also defined as follows:

There are many examples of this type of modification for production of complex of organic compound that have catalytic activity have of it chemically synthesized for example the myoglobin of which associated with oxygen, but it docs not have catalytic activity. This Bio-molecule with three complexes of ruthenium "ruthenium" carrier of the electron through the surface of the histidines components generate a complex that has the ability to reduce oxygen and the oxidation of the natural ascorbate.

The construction of DNA contributed significantly to the development to the stage of protein engineering to construct proteins that do not exist in nature. The technique has evolved to the point can modifies the gene by an engineering to change the protein in a predictable and have to improve some functional characteristics such as:

- No. transformation "turnover number".
- Static Km of substrate specific.
- Thermostability.
- temperature optimum.
- Stability and activity in non-aqueous solvents.
- Privacy of interaction and substrate "Specificity".
- Requirements of co- factors.
- Protease resistance.
- Allosteric regulation.
- Molecular weight and composition of the structural unit "Sub- unit structure".

And for engineering the protein molecule, it is clearly necessary to ensure a series of rules relating to major synthetic building blocks of proteins that recipe as desired. After seeing the structural composition of protein crystals, it is then possible to diagnose those areas in which it occurs possible modifications to improve the catalytic molecule, protein, and this is done to modify the sequence of amino acids in the protein.

Major modifications protein

The use of site-directed mutagenesis determined then what is aimed to, because the change in one base in the gene result in a change in the sequence of amino acids in the protein, which in turn improve the protein in question. Large modifications in proteins by removing the "delete" section mediated by enzymes or by the unequivocal chemical structure of part of the gene. In this way, the production of spare "klenow fragment" "DNA polymerase" free of analytical activity, also can add sequence of amino acids through docking to improve the stability of proteins made in E.

coli and finally can collect or part of a fusion gene or the whole of all or part of the other, thereby generating new proteins.

Determination the general features of the installation of the structural protein. Protein engineering based on the availability of information on the district and synthetic building blocks that are obtained from the methods of X-ray diffraction and nuclear magnetic resonance two- way "Two dimensional nuclear magnetic resonance NMR" and the latter is the alternative method in the future. Many researchers expect success in engineering of proteins "Protein Engineering", especially after the great progress which has been in embryonic technique, where each protein is produced by genetic conditions of its own machine of the cell consisting of enzymes when they become three characters of the genetic material and arranged in advance and checked that then wrap as a specimen to be specific proteins effectively.

When you know the rules that allow the protein to form belts wrapped can then change the genetic information of proteins and identified so that it works in another way as soon as a large and powerful grants stability, and thus can benefit economically from the proteins of the broad areas of application by micro- organisms and can be more clear: for example, improved production of proteins (new physical properties and functional).

Important notes that are related to protein engineering is to clarify the potential relationship of proteins, where the protein for example, a specimen 15- amino acid. There are 103×3 possible sequence of these acids is larger than the number of atoms that make up technical enzymes immobolized on board, the development of these enzymes are restricted or limited to a solid surface to be in constant contact with the foundation to which the article in the mobile phase "mobile phase". It is clear from this that there is a possibility to use the many pathways that retains its effectiveness.

Enzyme Technology

The biotechnology is considered as one of the technical life in science and engineering. It was one of the enzymatic technology trends that have grown with the technology of life, despite being preceded by technical life, keeping in mind that enzymes from an engineering standpoint is special case of the factors that have qualities such as privacy.

Bio- systems are used in critical periods in history to get the desired chemical conversions such as transformations of like milk to cheese and fermenting of liquids that contain sugar to alcohol, but such research trends have changed during the evolution of Biotechnology with the fact that these processes such as cheese, bread and alcohol industry still very important.

The history of enzymatic techniques started with the developments that have emerged a number of chemical transformations using the tissue of life, which include, for example hydrogen peroxide decomposition and degradation of starch to sugar and digestion of proteins.

Immobilized enzyme technology

At present, there are important industrial applications of immobilized enzyme technology represented by the following enzymes:
- Glucose isomerase.
- Aminoacylase.
- Penicillin acylase.
Lactase. Technical features of immobilized enzymes"
- Prevent the entry of the immobilized enzyme in the mobile phase.
- The product is characterized by being cleansed of the enzyme and does not accumulate.
- Using the enzyme for long.

The globe, despite the lack of clear understanding of the rules that govern protein engineering, but the equipment contribute to give some suggestions on how to achieve a stereo structure of the protein. In this area one can not expect for example bacterial cell to produce human that differs in form of human protein.

The latter has been "Immobilized" on the particles of silica. It is used to convert the lactose in whey to glucose and galactose.

Applications to include of immobilized in the future as follows:

• Use enzyme "Cholinesterase" for the purpose of pesticide detection "Pesticides" and watching the inhibition of this enzyme either by the method of electrical "Calorimetrically electrochemical" or by the color method.

• Other enzymes that may be used in the same method in order to detect toxic chemicals, the enzyme "Carbonic anhydrase" is very sensitive to low concentrations of chlorinated hydrocarbons from low- lying "Chlorinated hydrocarbon" and "Hexokinase" to "Chlordane".

• Immobilized diisopropyl phosphor fluoridate extracted from the nerve cells.

General aspects of enzymes immobilization

This process is intended as we mention it to determine kinetics of enzymes, as yell as cells that characterized by (desorption) on the surface such as fibers gels, etc., also can be used as phenomenon shooting accordingly.

Advantages of the immobilization process are the followings:

• Finding the status of enzymes similar to those found within cells and tissues.

• Prolonging the period of use and has repeatedly given to the survival of catalytic activity and stability.

• Use appropriate concentrations and may be high for the purpose of increasing the speed of the reaction, given the focus to fit with the speed in specific circumstances of the reaction.

• Contributing of the immobilization process to facilitate the purification process of related to products of reaction.

- The use of multiple systems from the fermentation (continuous and open).
- Reducing energy consumption and cost.

The immobilization methods are numerous, including:

- Chemical methods: they are similar to affinity chromatography such as use the covalent and casual.

- Physical methods: such as packaging inside a capsule adsorption and shooting.

As for choosing the appropriate method to be immobilized it is determined according to the specific bases represented by measurement of activity and stability. So it must be taken into account the business side which means less expense. And choose the easiest method because they are all tough and stay away from hazardous substances to human health, and the technical side is important in the selection process since there is a special mechanical pressure during the operation.

The immobilization cells vary from cell since it is being more of enzymatic system builders with the installation of diverse chemical content, therefore, requires that the appropriate modalities, simple and stay away from these that require to use extreme circumstances. It also requires that to taken into account the number of cells to be immobilized so the method must be convenient and linking cells are good and avoid the use of hazardous materials. The characters of the immobilized cells are numerous advantages including the use of small amounts of carbon and energy sources and re-use of cells, so it is possible separate the growth phase from production phase, where it is possible control the fermentation. Immobilization depends on the type of cells, microbial cell reduce the size of the manufacturing process and thus reduce the cost of the production process. The Eukaryotic cells which are characterized as specialized capable of limited division of which are specific plant or animal cells and preferred to be immobilized, particularly those that are separated as any single and are generally used for the purpose of the immobilization of adsorbed on the hollow fiber.

Human Genome Project

The initiative of (HGP) came the first time in 1988 and aimed at finding the sites of some 100000 human gene in DNA and the (HGP) expresses 24 pairs of the human of chromosomes, is turning into information content when it follow the sequence of rules need to be resolved based on computer science, mathematics, statistics and experimental sciences.

The computer science often provided in contributions in programs and solutions that are characterized by the skills that led to the invention of language access code information described the performs a particular order and provides methods to describe complex biological processes by the number of code rather than their natural language with hundreds of pages. Then the researchers and observers said that the twentieth century be the century of biology and analytical power resulting from the HGP that will explain drastically all life and medical research as it was:

- Research on the nature of genomes and the nature of the composition and organization of various scientific institutions.
- Acceleration in the implementation of the project, which was planned to complete within 15 years of technical progress, but then shortened the time to ten years   .The search of the genome to find the type of information or material contained in the communication as well as identifying the sequence about three billion chemical bases.
- Study the nature of the information stored in the computer and the evolution of elaborate by efficient techniques of and sequence evolution in the tools that contribute to the analysis of information.
- Study the effects to be set in the community and to what extent this

                                                                                    can

be achieved, and the type of response.

In spite of all reported studies and research conducted by methods and techniques in various vital information as well as numerous writings and published in this area, there are still many other fields and various study and research. Some of these fields has not been touched so far in the country, especially the human genome projects, and areas to attract the attention of researchers, but it's mostly a few problems, mostly dealing with partial or subsidiary.

Bio-engineering

There are a number of scientific developments resulted from the diving in the world of molecules to push medicine forward through the discovery of technical of recombinant DNA (engineering life) and this new knowledge has led to the understanding of the causes of the disease that has eluded science until now, and thus to find new treatments to them. Engineering of life had an impact on medicine borders these have become easier with the forgotten youth of this important scientific field.

The reality is that James Watson and Francis Crick did not reach a structural installation with a double helix molecule of DNA. And then it was identified the gene (genes), which manages the production of individual proteins, and then we obtained the tools of partial strong, and in the early seventies researchers began snapped genes of the DNA. One of the species and planting it in DNA another kind for the manufacture of new molecules and in a few years researchers were able to transfer these genes and to produce objects that are within during the eighties and became a human gene transfer to many microscopic organisms and bacteria turning them into factories for medically useful proteins.

After it has been cloned of human genes in the micro-organisms for a number of hormones, including growth hormones and insulin in human as well as bacteria many of the genes responsible for human proteins with diagnostic value was produced at the level of marketing. It is noteworthy that human insulin is derived from living with diabetes, and also for the development of techniques for the production of antibodies "monoclonal antibodies".

Many applications, there is a steady increase in the use of enzymes in the diagnosis and treatment as well as in planting (farming) tissues and cells,

"Tissue and cell transplantation" and that the development of engineering of life is still in the young stage, but there have major impacts on medicine and industry is synergy between electronic systems, electrical and life-component electrons so-called life "Bioelectronics" and electrochemistry of life "Bioelectrochemistry". Then there have been the following design of a number of devices depending on what is stated in the above examples include "Glucose monitors" for the purposes of medical sensors and nerve gases for medical purposes and sensors nerve gases "Nerve gas sensors" to military uses.

Based on sensors that have been most developed in the present time to reveal the exact products enzymatic activity mediated by the traditional pole "Conventional" where is the install (restricted) "Immobilization" new approaches that lead to devices with more sensitivity that depends on the movement of electrons between the direct-polarization and the redox centers protein "Protein redox centers" In brief the enzymes, which is based on the sensor depend on the medical sensor "Glucose sensor" and other sensors that measure chemicals in blood such as immune sensors include the electronic life "Bioelectronic immunosensors", which was commercially manufactured during the current decade, are measured in a large number of materials in the fluid of life, causing a revolution in the diagnosis, in addition to the incremental progress that has been happening as a result the development of a wide range of models "Sensors", which depends on the synergy between micro-organisms substantiated grants stability, "Immobilized and Stabilized". Finally, various data indicate that the microbiology of life through the engineering involved in the medical field in the production:

- Antibiotics.
- Vitamins.
- Nucleotides.
- Hormones.
- Enzymes.
- Vaccines.
- Antibodies.

The progress that accompanied the engineering of life has affected in particular the daily practice of doctors, because of the speed that accompanied the evolution of knowledge and techniques in the laboratory and hence to the industrial production and then patient care. The expression of human insulin gene in bacteria E. coli, for example, has been studied in 1979 and that this insulin, with the original engineering-

life of "Recombinant DNA" has been tested by volunteers with non-diabetes "non- diabetic" in 1950 and clinical trials that have been in patients with diabetes began in 1981.

The attention of most doctors on the applications of modern life engineering in medicine, which tend to be very important in areas which have helped to revolutionize the diagnosis, treatment and understanding of many diseases, and examples of this therapeutically important protein, which was manufactured by engineered mediated microbiologist, microbiology, applications of single origin "Monoclonal antibodies", enzymes and others that arise out of uniform origin from lymphoid cells, where used in:

- Treatment of cancer.
- Diagnosis of many diseases.

Pharmaceutical industry, pharmaceutical companies have been choosing some clinically significant produced cheaply, such as insulin, which was previously mentioned, and which treats patients with diabetes and extensive use of interferon for the treatment of many diseases, including cancer.

The Bio-engineering worked towards a second method by increasing the secretion of microbiology by called anti- life penicillin produced in fungi, and the third trend in the medical field that is the development of drugs already in the nature and turn them into centers of drugs more effectively.

Containing anti-bacterial drugs that have contributed to engineering life and developed vaccines, hormones, vitamins and antibiotics and life for the purpose of producing these materials from micro-organisms after it was restricted to human and animal cells.

Hormones are the most advanced in terms of the accuracy of the technique used and the large economic returns through the engineering of life and led to great successes through the production of materials likes of the hormones which are stimulated, and stimulating the flesh wounds and the growth of the affected nerves that affect the sense of pain. The success of engineering in the provision of life-hormones of the study and treatment has become a boom due to technical difficulties in extraction, which vaccine and growth hormones as well as the instigator of the secretion of pituitary adreno "ACTH" used to treat infections and diseases is used to treat wounds, burns, and stunting and release thyroid hormones pituitary as well as insulin used to treat diabetes, where possible transmission of their genes to bacteria.

Production of hormones is mediated by microbiology research center in the fields of engineering life in general and genetic engineering in particular, where microbiologists used to convert steroids and the production of hormones from the human body can not produce in sufficient quantities. Then it was grown in importance after the custom of cortisone and its derivatives and their effective role in the treatment of arthritis, which draw many medical companies of steroids from plants, animals and chemical methods of trying to turn them into other steroid prescriptions.

The methods of microbiology steroids is turning quickly but with less degree, there is in the addition of specialized microorganisms capable converting steroids quickly. There is also the addition of specialized microorganisms is added hydroxyl group of any carbon atom present in the steroid.

There are also some working to add hydrogen to steroids or withdrawal of hydrogen or oxidation or separate pools of chemical side effects. Using growth hormone that is released from the pituitary gland for the treatment of dwarfism find the hormone extracted from the animals be in a non-pure from, but according the production of this hormone is preferred to be extracted from microbiology such as the production from the bacteria E. coli after treatment genetically.

The plant hormones have been possible to produce from fungi, especially those produced from rice, as it is known that plant hormones industry is still expensive despite their limitations. In addition, there are a large number of proteins found in the blood such as the factors that contribute to coagulation missing by patients with hemorrhage as well as the albumin found in serum. These materials have been contributed to the development of production by engineering life in medicine (drugs).

The pharmaceutical industry, which includes anti-bacterial drugs, vitamins, vaccines and hormones of the biggest industries that relied on engineering techniques of life for the purpose of producing these materials from microbiology.

Medical applications of Bio-engineering

There are many faces, can be addressed when studying the medical applications of bio- engineering after the gene was designed, including:

Production of therapeutic: include hormones, such as somatostatin insulin, interferon and anti- biotic, where it was initially isolate the hormone somatostain for regulating secretion of growth hormone from the pituitary gland in the traditional way that requires half a million sheep brains to produce 5-10 mg of this material.

Treatment of many of the genetic diseases: the treatment of many genetic diseases is possible to treat many genetic diseases due to loss of protein production remedying these proteins from bacteria, and the examples of this case the planting and production of large amounts of genes to produce hemoglobin, which decreases in "Thalassemia" through the introduction of genes responsible for hemoglobin the patient's bone marrow, and then returned the cells to the patient.

Diagnosis of a number of diseases before birth: the fetus diagnosed in the prenatal stage, through the identifying the defects in a specific gene that causes the disease, such as some "Gamma- Globuinemia" and the disease lest Nhin as well as Tay- Sachs "Tay- Sachs".

There has been progress in some areas of medical engineering technology due to the recombinant DNA such as "cloning" the human insulin gene as well as growth hormone and its expression in bacteria that has been marketing of human insulin derived from microbiology and used for the treatment of patients with diabetes in addition to:

-       Production of interferon by a large clone human genes in microorganisms.
-       The development of production techniques and monoclonal antibodies and their uses.
-       The increase in the use of enzymes for the diagnosis and treatment in instilling the cells and tissues "tissue and cell transplantation".
-       Treatment of many diseases of genetic mediation by protein that being lost, which can be mediated by production of bacteria.
-       Diagnosis of diseases before birth by identifying the defect in a gene or several genes.

Turning to the relationship between engineering, medicine, is taken into account the following things:

• Mutant cells and the cells unmodified organisms and their products such as antibiotic cellular life and plants, as well as other life transitions "Bioconversions".
• Modified cells "Modified cells" and their products to ensure that objects Monoclonal "Monoclonal antibodies" of the following uses:
-       Immunological Studies.
-       Immunohistochemistry.
-       Tissue typing for trans- plantation.
-       Diagnosis and monitoring of malignancy.

- Preparation of medicinal products with a "Prepartion of medically important products".

- Recombinant DNA technology and its use for the production of insulin, interferon and growth hormone and vaccines "Vaccines" and enzymes.

- The application of Bio-engineering techniques of molecular

Genetics and techniques diagnosis recombinant DNA in the diagnosis and (pathological) human disease:

- Patriarchal diagnosis of genetic diseases.

- Effects of genetic diseases on the specie disease.

- Features of the future.

It is believed to that Bio- engineering represented by "Clinical biotechnology" has begun in the application management and industrial production of penicillin in 1940 that the success of the full insulin has created a growing demand for medicine (drugs).

The production of penicillin by fermentation and used in the treatment of diseases using the Bio-engineering problems that has been accompanied by the emergence of side effects and put some Bio- engineering solutions, and the problem of production has been developed through genetic improvement producing strains and control the components of the center other conditions contribute to the process of fermentation.

Bio-engineering and cancer

Bio- engineering has succeeded results in the field of cancer better than other diseases, as shown in the eighties that the main thing vs. cancer is a change in the genes (genes) from an engineering standpoint.

It was clear from the following entries in the relationship between the Bio- engineering and cancer.

Through analysis of a group of viruses called regressive "Retroviruses", which cause cancer in animals, as a number of these viruses carrying cancer-causing genes or tumor genes "Oncogenes". It appears that the retroviruses that cause cancer have been captured from the normal gene, cell; or one animal that is made part of their own genetic material. The retroviral infection of new cells in the later planted with genetic material, leading to the transformation of healthy cells into cancerous cells.

The researchers show that DNA extracted from human tumors can shift the cancer cells to cancer cells in test tubes. Or that a specific gene in a human cell that can transform sound cell into the tumor cell and a tumor-causing gene for bladder cancer in humans.
.

The gene tumor is often due to the mutant or increase in production and there is general consensus about the fact that any of the original tumor gene mutations may be some inherited mutations.

The studies of funmor contribute to inherited breast or ovarian caner, the physician may be able to use that gene to assess the patient's condition and prospects and to provide more effective treatment for patients who have multiple copies of inherited suspicious. Harold Varmus and Michael Bishop has concluded that "Lancogen" the legacies of the genes responsible for causing cancer.

Bio-Engineering and AIDS

To understand the relationship between Bio- engineering and the AIDS requires a study of the topic in two cases:

How should the immune system to destroy virus: the defense forces resulting from the immune system to attack multi- directional and of different media for the virus (a specific target) to:
-       Phagocyte and other cells relevant to specific viral antibodies are chewing.
-       These cells installed in the grooves on proteins known as antigens of human white blood cells.
-       Construction immune complexes on the surface of cells identified by a type of white blood cells (T-help) "Helper T".

- The recipients are on the T- cells help identify the peptide superficial "epitopr", associated with divide, and secrete small proteins that stimulate and activate T-cells and the toxic or lethal trait.

- The killer T cells directly attack infected cells and fragmentation of viral particles and peptides associated with molecules of antigens of human white blood cells, when identified by toxic T cells by antigenic recipients on the surface of infected cells and destroy them by producing more of them.

- The B- cells recognize the antigen norepinephrine viral surfaces as a prelude to their destruction.

- Immune response and the virus "HIV" contribute the immune steps in defense against the virus "HIV", where they are:

▪ Invasion of the virus of T- lymphocytes and cells assistance, followed by cloning and increase the virus and help decrease the number of cells, death, and loss of infected T cells.

▪ Launch of viral particles from the cell membrane of T cells after being wounded by the T cells and B- toxic responses to be dispatched a strong defense which resulted in killing infected cells, viruses, and thus is determined by the breeding assistance and reference cells to a normal level.

▪ A high level of virus gradually with the decline in the number of cells to help patients and reflects the so- called phase of AIDS when the number of cells less than 150 assistance cell in the blood followed by a rise in the level of virus with the decline of the immune system.

Monoclonal antibodies

The areas of application for the production of these antibodies where the potential for many therapeutic and diagnostic enormous, including:

• Treatment of patients with leukemia and production of specific antibody alien objects on the cancerous blood cells, leading to the union of antibodies with and removed from the bloodstream.

• Accepting the objects of a transplanted organ which are used Monoclonal antibodies or clone in the development of the body accept a transplanted organ such as the kidney.

• Birth control through private industry specific antibody to proteins found in human sperm.

- Determining the sex of the fetus through a special antibody to sperm of own unwanted sex.

- Models are highly sensitive and privacy are being used as opposites, and a single origin and widely high sensitivity and privacy in early screening for malignant tumors by using specific proteins associated antigen and the presence of tumor presence.

- Determining the levels of hormones in the body and used Monoclonal antibodies to determine the levels of hormones in the body and determine the effectiveness of the glands.

- Search for the presence of some drugs in the body tissue and blood used Monoclonal antibodies in the search for the presence of some drugs in the body tissue and blood to prevent the occurrence of cases of poisoning or addiction.

- Diagnosis of crimes using Monoclonal antibodies in the search also in the diagnosis of crimes. The food industry also used Monoclonal antibodies in the field of food industries, especially in the diagnosis and determination of the purity of food, processed meat, and free of

-

unwanted substances and preventing fraud in this area.

Of the significant developments that have taken place for Immunology and molecular biology and biochemistry and the discovery of antibodies and the creation of a single origin "The Monoclonal antibodies" is characterized by privacy "Specificity" and sustainability of production,

"Immortality" huge quantities "Large ruantities" and high purity "High Purity" for periods of a very long time.

However, these antibodies Monoclonal antibodies created by the multiple origin (clone) the molecular composition and effectiveness. Studies have shown that the use and applications of antibodies only be successful to detect very small quantities of tumor functions that can be used in early diagnosis of many tumors and by diagnosing the effectiveness of these antibodies could be argued that a large proportion of blood diseases can be categorized.

The advantage of imaging the immune flashlight as we have mentioned that the blue single antibodies prepared in the body of a patient associated antigen, surface of cancer cells without other cells and sputtering when labeling these antibodies with radioactive isotope, it can locate the radioactive iodine, for example by gamma cameras and thus can be located and the size of cancerous tumors, including colon, ovarian and skin cancer.

The unilateral clone in addressing some of the tumors where it can be linked to medicine as well as radioactive materials to these antibodies, such as chronic leukemia and thyroid cancer lymphoma and colon cancer has been found that these antibodies injected intravenously is grappling with the tumor cells and selectively and is disposed of, where became can direct these drugs directly to tumors by linking them to the catalytic antibodies to these tumors. Used monoclonal antibodies to treat cancer when there is a high toxic concentrations in the tumor. It could also be linked Monoclonal antibodies radioactive isotope and alive in the body of a cancer patient at which time the radioactive material to the site of the tumor and therefore within the cancer cells and it crashed. There are many researches addressing the use of monoclonal antibodies in the early diagnosis of the body rejecting the case of the tissues and the transplanted organs as well as a lot of studies on the use of these antibodies in the treatment of the case of rejection.

Some applications objects Monoclonal

Improving the sensitivity of the current immune for tests or tests new Histocompatibility
Fibronnectin
Blood groups Antigens
Sperm antigen
Interleukins IL
Interferons
Progesterone gastrin
Blood   clotting   factors
Estrogen
Human growth hormone

Monoclonal antibodies has clear impact and important role in clinical medicine before developing the "hybridoma technology", which provides heterogeneous objects "Homogenous antibodies". The research carried out by each of the "Kohler & Milstein" in the early seventies, created a method used for the manufacture of the anti body homogenized with a quantity of non-specific proliferation applied at large.

The researchers "Kohler" and Mlesstin have participated in the production of monoclonal antibodies, which is derived from specific tissue culture which is called hybridoma, where the latter's has the ability to produce one type of antibodies but does not produce more. This is done by crossbreeding or mating types of cells, the first is produce the antibody and the second for the growth of cancer cells have the ability to reproduce. And then treated with hybrid that has to be the formation of antibodies, where antibodies are produced for this body alone, and perhaps it carries the qualities of cancer, the production of antibodies is very large quantities. It is possible in the light of the use of a composition for the manufacture of an unknown antigen monoclonal each part, and then used these antibodies to probe the chemical composition of the real knowledge of the unknown substance.

Monoclonal antibody can be used for treating patients with cancer of the blood through the manufacture of these antibodies is specific to the alien objects on the cancerous blood cells united for the purpose of removal from the blood stream, and used these antibodies for early detection of the presence of tumor cells through the tests that require purity too high to measure the presence of proteins associated with its existence of these tumors and their locations in particular antigen- mediated tumor.

These antibodies are used in determining the levels of hormones in the body and to determine the endocrine events are also used in the search for the presence of certain drugs in the blood and tissues because of the poisoning have also been introduced in the diagnosis of bacteria in the development of the transfer of the body of a transplanted organ, in particular kidney .

Preparation of medically important products

The use of monoclonal antibodies the purpose of purification and preparation of medical materials which is represented by the task done by the "Secher, Burks" and that covalently bound to anti- body unilateral origin against assigned to drive in "Sepharose" and therefore can refine 5000 times.

There are a number of the production of insulin- led company "Eli Lilly & Co", which used the Bio- engineering, including recombinant DNA technique as a base for the manufacture of human insulin. The production process has been carried out by "Lilly" in cooperation with "Genentech Inc." According to the following steps:

• Determination the sequence of DNA From the known sequence of amino acids m insulin.

• Chemical structure of genes for the series "A" and Series "B" of insulin, contains each and every one of them in the codon methionine at the end "O".

• Each gene enter in 2 mentioned in the beta- 2 gene "B- galactosidase" of the plasmids, which are the same within the" E. coil".

• Because of the fact that the bacteria had grown in the medium that contains the galactose and not glucose that urges enzyme B- galatosidase and then with a series of insulin "A" or "B" linked with methionine.

• After the breakdown of bacteria, treated with cyanogens bromide "CNBr", that breakdown the proteins at the site where the

• methionin is present.

• Purification of the two strings "A& B" and then returned to their union with the natural production of insulin, the two strings.

The bacteria do not carry enzymes that change the pro insulin to insulin through manufacture the lofty strings "A, B" in the bacteria followed by purification separate chains bilateral, ties with the sulfide.

It is noteworthy that human insulin that is produced by E. coli was tested by healthy human volunteers and with diabetes (and there were no adverse that there is capacity similar to purified pork or with decrease the blood glucose when injected under the skin or injected inside the vein.

The trials have been carried out on human and compared with animal insulin producer from the pancreas of pork produced then was shown similar effects. This was in the hospital, "Guys" in London and Osaka in Japan.

The department of food and drugs in the United States the U.S. approved marketing of insulin produced by the micro- organism, which called the "Humalin", also got the same thing by Britain in the same year 1982.

Growth hormones

These hormones are used in the treatment of growth disorders in children, dwarfism some cases of infertility in women and because of the high cost of treatment using these hormones and the difficulty of obtaining and the need for a large number of the pituitary gland the high cost of the hormone and the likelihood of infection with viruses and expected such as uses future growth of the tissues, and the flesh wounds after surgical operations the flesh of fractures and the assist once in the treatment o burns, sores and the used in the study of malignant diseases.

Researchers have made efforts to extract it and its production in bacteria by bioengineering after the successful transfer of genes responsible for hormone production from human cells to bacteria which was done in 1979.

The length of human growth hormone, 191 amino acid with a molecular weight of 2200 excreted by the pituitary gland, secreted of from the front gland of the longitudinal growth of the structure which means the isolation from the pituitary gland.

The pharmaceutical kebi have cooperated with the company "Genetech" for the production of growth hormone from E. coli using Bio- engineering of life recombinant DNA techniques.

The technical difficulties have been overcome in the collection of pituitary glands as well as the cost that accompany and other problems, including lack of access to one type of the hormone, but a mixture composed of several different forms in the installation of structural and molecular weight than did patients with antibodies inhibiting their production.

Interferon

The interferon is extracted from the cells infected by virus and studies have shown that these cells, the immune system stimulates the production of this hormone during the infection by virus.

Then overlap with the later injuries to the work therefore it is called "Interferon" used to treat the casualties.

The cost of purification of interferon is very high and extracted from white blood cells, and the other efforts made to develop the production method for human interferon non-blood through tissue culture (artificially), and then methods developed to produce interferon from bacteria, it is done successfully in 1980 and it became clear that there several species produce a number of genes.

Interferon has been isolated in 1957 and considered at that time the first line of defense against attack by viruses, used to treat many viruses diseases and including:
- Cold.
- Hepatitis.
- Cancer Diseases.

Because of the interferon ability to prevent abnormal complications of the cells studies have shown that the immune system of these cells motivated during virus infection leading to secrete very active material is to overlap with the later injuries to the work.

The interferon's a family of proteins, discovered as a result of infection which flows into the cells by viruses "Virally infected ceils" are characterized by the following characteristics:
- Antiviral in other cells.
- Inhibition of "Cellular proliferation" "anti cancer drug".

- Internalization of the immune system "Modulation of the immune system".

It is possible to re-classify the life can be interferon's to the following types:
- α- alpha- interferon leucocyte "α leucocyte interferon".
- β- beta, interferon cell Fiber "fibroblast interferon".
- Gama of lymphocytes, immune interferon "immune interferon", "Lymphocytes T".

Leucocyte interferon reduces the spread of a vesicular composition "vesical formation" and responds to fear of infection of the liver "Hepatits B" as well as in various malignancies "malignancies" such as breast cancer spread "metastatic breast cancer" and non- Hodgkin's lymphoma "Non-Hodgkin's lymphoma" and osteoma flesh "Osteosarcoma" and malignant melanoma "Malignant melanoma".

Research has shown that there are about 20 types of interferon, which produces a number of genes mediated for the purpose of genes engineering more effective against viruses or against tumors.

Production of interferon's

- Using 50000 liters of human blood, produced from 0.1 g of pure interferon for the treatment of acute viral diseases.
- A culture producing cells and white blood cells of some healthy donors and encouraging the virus "Sandi Virus" for 24 hours and then isolated interferon-mediated by centrifuge in the Central Laboratory of Public Health in Helsinki, France and the United States.
- DNA recombinant technology.

Biotechnology

The concepts of traditional genetics known for thousands of years and then evolved into the development of Mendel's laws, famous, and then changed, according to the progress of various technologies and discoveries.

154

The increased in gene progress in terms of chemicals, then was reached as to how the work of the gene at the molecular level with the adoption of methods of biochemistry, rather than traditional methods in the interpretation of genetics, which paved the way to the evolution of the concept of genetic engineering.

The high and low living organism from units can only be seen with a microscope, a cell, which contains the kernel and the last, which includes chromatin materials that turn into chromosomes (chromosomes). Studies show that both the egg and sperm contain half the number of chromosomes (chromosomes) and therefore half the number of genes in human egg contains, 23 chromosomes. Chromosomes chemically composed of proteins and the "DNA". Studies have indicated that the gene is made up of sections from the chemical DNA. The latter consists of chemically according to the double helix model of Watson and Kirk.

That has been proven that DNA is the genetic material according to the following:
• The amount of DNA constant in all cells of the individual, regardless of the quality of tissue that make up the member.
• The DNA ability to configure a mirror image of himself during the division.
• The DNA characterized to contain all the genetic information in the order of succession of base nitrogen.

Genetic engineering
Genetic engineering is intertwined with the vitality and technology based on several basic sciences such as cell science, genetics, biochemistry, physics, and others.

The content of genetic engineering the human ability to control the mechanisms of gene transfer from one cell to another and how to express them within the cell for the future.

To understand the genetic engineering practice to be done the following:

- Isolation of DNA of the object which is meant the transfer of its genetic material.

- Cut the DNA to the sections that each end section to a particular gene.

- Identify the gene required between these parties.

- Ensure the presence of a carrier "vector" suitable for gene transferred in order to carry the gene of the object to the donor organism.

Discoveries that paved the way to genetic engineering:

- Carrier "vector".

- Types of bacteria contain a small chromosome called "Plasmid".

- Restriction enzymes DNA, you cut it off at specific sites.

- Ligases close the gap left by the restriction enzyme.

- Select a succession bases in the DNA Sequencing.

- Synthesis of pieces of DNA "Oligonucleotide synthesis" for the purpose of identifying genes within the cell for the purpose of diagnosis of many genetic diseases and led to begin the implementation of the Human Genoma Project.

- Using "Probe" in the processes determining the existence of gene and diagnosis and genetic makeup of the individual.

The genetic engineering since its birth in the seventies of this century a lot of fear, it is double-edged sword usable for good or evil and see the use in the prevention of disease or treatment, whether genetic surgery that change the genes with other as well as another gene in the filing of another object to obtain large quantities of secretion this gene for use as a drug for some diseases. After the success of the possibility of transferring genes from one cell to another, there were some concerns, including the following:

- The possibility of introducing genes that are synthesized toxic material within the cells of bacteria and make them so harmful effect.

- The introduction of parts of DNA Tumor Virus in another virus bacteria.

- Disable the genetic diversity, where the plants or animals that were subject to genetic engineering are usually homogeneous, making it vulnerable to bacterial and viral diseases and others.

Scientists played down such fears, the development of standards and controls to reduce the risk of manipulation of genes and these terms:

The issue of genetic engineering, genetic since its birth in the seventies of the last century a lot of fear, they double- edged sword usable for good or evil and see the use in the prevention of disease or treatment, whether surgery, genetic change since intra gene in the cells of the patient and the gene in the filing of another object for large quantities of secretion of this gene for use as a medicine for certain diseases while preventing the use of genetic engineering on sex cells "Germ Cells" for the legitimacy of the dangers.

And taking into account that it also may not be used in genetic engineering purposes, evil and aggressive, or to overcome the genetic barrier between the different races of creatures in order to create objects out of shape, mixed curiosity.

It is not permissible use of genetic engineering policy to alter the genetic structure in so- called improvement of the human race and any attempt to tamper with the genetic character or human intervention suited to individual responsibility is legally.

The prospects and risks of genetic engineering (genetic) after the success of the possibility of transferring genes from one cell to another, there were some concerns, including the following:

- The possibility of introducing genes are synthesized toxic substances within the cells of bacteria and make them so harmful effect.

- The introduction of parts of DNA tumor virus in another virus in bacteria, with the spread of these viruses and bacteria spread of the -disease.

- Loss or interruption of genetic diversity, as the plants or animals that were subject to genetic engineering are usually homogeneous, making it vulnerable to bacterial and viral diseases and others.

- It is known that human intestine contains different types of bacteria the bacteria can thus be dealt with genetic engineering techniques to live in the human intestine and increase the chances of the spread of diseases and epidemics.

However, the scientists played down such fears, the development of standards and controls to reduce the risk of manipulation of genes, some of these precautions: Controls for the design of laboratories and security measures to prevent the spread or leakage of bacteria and viruses treatment.

Although the benefits of biotechnology in many areas such as medicine, agriculture, industry and conservation of the environment there is increasing data show that diversity of thought began to intervene in the subjects continued to reduce the time from the jurisdiction of social thought and reason. The evaluation process of biotechnology has not yet started, does not believe that the limited studies on the impact of bad to launch microorganisms genetically engineered has no real value. These studies resulting from the overlap of politics in science and nothing to do with the problems faced by our communities.

Protein map "Proteome"

The term "Proteome" which appeared in 1994 to the total pool of proteins present in each cell type the amount of a hundred trillion in each individual and the total proteins produced by cells of the body during different life stages.

After discovering the human genome, which includes (full content of genes (genes) in the amount of the 34 thousands people only, and not one hundred thousand. I think scientists for a long time) and also all the genes inherent in the cells of the body at the present time highlights the important question, what the protein content of these cells.

The type of each protein has to be known as a result of these cells and what function each protein and then what order of these proteins. Asking this question came after attrition rationale the concept of the genome and its consumption and is not enough to know as responsible for stimulating cells to produce the kinds of protein, but only requires the knowledge of the situation in its entirety in routine cases of disease and natural and in accordance with these questions and answers on the back of proteome.

Proteome contains information more complicated and the secrets of the genome is more dangerous than those found in the genome and extensive knowledge and synthetic for more than a million different types of proteins. The concept of proteome is known later human proteome is doing now by scientists and they hope that these will be the beginning of the main achievements under this project, despite the severe difficulties faced by these scientists, in excess of those related to the human genome.

The analysis (cell proteome), reached some of the researchers in 2000 to build automated device called the molecular scanner "Molecular Scanner", which is carried out by measuring by mass spectrometer, from which tens of thousands of known proteins in a single day, and at the speed of more than ten times what was known before.

These researchers also managed to build a million boosted the analysis of protein per day to build bigger infrastructure database proteomics mankind. The draft of human proteome or other whereupon many of the laboratories and big budgets and international companies, different research directions, the analysis of three-dimensional protein structure and interactions between proteins, which performed many of the key characteristics of the human proteome would pay off represented by the following:

- Specification of fungus or yeast proteome with a single- cell, the first that has been done in the world of proteome.
- This project change from how to design drugs in the near future.
- The appearance of the so-called science and technology human proteome, which will focus on the conversion of most of the drugs manufactured by genetic engineering and biotechnology.

Genetically Modified organisms (GMO)

Biotechnology, which contains the processes of nucleic acid technology and molecular biology to separate the specific gene from one organism and transferred to a particular object of another district called gene transfer technology, "Transgenic" or may be called the genetic change or genetic modification "Gene modification" and called on the living modified organisms has been applied this technique recently on agricultural crops in the recent developments in genetics, which is also hot topics in it. A number of genetic modifications on some common food organisms, the addition of a specific gene or several genes, for example when carrying out genetic modification of wheat plant is usually a small percentage due to the fact that this plant has about 80.000 genes.

The process of genetic modification is possible in practice so as not to become genetically modified plant to another object or to plant malicious, but maintains the general attributes of the amendment with relative injury. According to some voices of opposition to the process of genetic modification, that could cause damage to humans and the environment, including poisoning or allergies.

The number of countries including the United States of America, Canada and China that will produce genetically modified crops, including soybeans, corn, flax, beets, potatoes in different proportions. From a technical point of view alone, there are a number of benefits of genetic modification of crops to convert to regular crops resistant to pesticides and weed, disease and insects and reduce pesticide use and increase productivity and improve the nutritional value of crops, and make it more a shift of the circumstances, including salinity, drought and an increase in the quality of the crops for use in food as well as in withstand the transport and storage and make crops resistant to pesticides, insects or insect resistance or both groups.

The genetic modification was still in use in plant breeding has a significant impact in providing food for humans and methods that have been used traditionally to improve the crops, but they are not specific or accurate results of modern genetic modification in which the change is unknown in most cases, things such as crops and breeding plant breeding and mutations.

Advanced sciences and technologies

That the era of advanced technologies "High Technologies" or high-technology "Super Technologies" in which we lived the last three decades of the twentieth century, the era in which we do not know how many decades it will take, representing a number of scientific areas and new technology comes on top of these technologies, laser and fiber-optic and space technology, new materials, pharmaceuticals, chemicals, minute nanotechnology, and finally biotechnology and genetic engineering.

The forthcoming technical applications that are difficult to know the extent of today and its impact on humanity can be viewed as the era of advanced technologies as the following day when mankind as a whole interconnected network giant relies on a wide range of communications satellites such as radio waves and X-ray laser, so that every part of the ground contact one of the satellites in the moment and will be available electricity in remote areas with farms, genetically engineered to convert sunlight into carbon and then to the crude stream, and can then run all the equipment and facilities, communications equipment, including satellite and the Internet.

The future applications of these technologies will be radical changes in the forms of life activities and practices relevant to the interests of individuals, groups and the process of coordination between these advanced technologies is a strategic way to bring about a surge in operations research and industrial beginnings began to appear, for example, a draft genome and bioinformatics. We will try in this article and subsequent articles offer examples of advanced technologies

Nanotechnology, the technology that deals nanometer scale (metric unit of measurement), which is manipulating atoms for the manufacture of automatic equipment and information does not extend far a handful of atoms, and then anything can be made by micro- physics is quite different. The first part of the term to reflect the unit of measure (nano= 190^9 meters) the term nanotechnology in Engines of Creation written in 1986 and said the possibility of seeing the future of the armies of machinery hidden carry oxygen and nutrients and waste, and manufacture of atomic- sized machines called complexes pads that hold individual atoms. It was found that the control of the maize one and move freely and easily from attributes of nanotechnology.

This technique showed high density in the form of recent innovations in many of the global scientific publications. Including the mutations responsible for many genetic diseases and therefore provide in the future and relevance of information is indicated task to determine the organism is an information system for the manufacture of proteins and other compounds in spite of the inability to resolve the issue of structural triangular shape of proteins, despite the existence of mathematical models serve the purpose.

A fruitful field of nanotechnology research in many parts of the world and in government labs, commercial and academic, have emerged, according to the products on this technique such as sewing pants of fiber and manufacture of precision tennis balls retain flexibility. In the near future, computers will appear smaller tubes made of carbon atom chips represent the atomic scale wires and high strength to build elevator to space, and plants that will manufacture computers minute integrated directly with the human brain to increase intelligence.

It is clear from the examples mentioned that nanotechnology is a technology that will change very little minutes every aspect of human life and giving people the ability to control the material, and this technique is the most important applications of medical treatment of human beings through the introduction of precision instruments within the cells to repair infected objects from within or for the diagnosis of patients as well as some developments on the mechanisms of control cells. The first medical use of this technology has been developed a device implanted in the body that may manufacture.

The energy comes from bio-fuels in the cell and the purpose of this engine is to integrate machines in living systems fully, and use some of the scholars of this technology to produce nano-bombs to kill cancer cells. A team of other customers of this technology to produce nano- bombs to kill cancer cells. A team of scientists of any other industry, the crew of siliceous teeth not larger than the size of the cell that can swallow the red blood cells and re-launched into the bloodstream, either antibiotics nano-particles "Nano biotics", which new types of antibiotics contribute to solving the problem of resistance of some types of bacteria to drugs, as well as the modified bacteria organisms, are converging nano- tube micro- rings 2.5 nm diameter amino acids and small hole walls of bacteria are infectious.

Researchers believe that the future of medicine is moving towards nanotechnology that will change medicine, as future devices that will work within the human body to diagnose many diseases and treatment. Russian scientists in the field of quantitative light and laser physics to reach a new discovery has been called the needles agency, a new type of X- ray beam, or special characteristics, as containing the elements of nano- any electronic material on subatomic particles that do not exceed the measurements of nanometer dimensions .

It has also developed the first computer chip companies auction that could contribute to increase the power of computers and a reduction, while reducing the amount of energy consumed by the chip is composed of cylindrical molecules of carbon atoms in diameter than a billion to a part of the linker carbon (smaller than a hair a hundred thousand times).

Femto means the number 15 and the chemistry of femto, to understand the reasons that lead to some chemical reactions without the other, one of the achievements made at the end of the twentieth century and the efforts of the world that have emerged Ahmed Zewail, who won the Nobel Prize in Chemistry in 1999 and showing the possibility of seeing how to move the atoms within molecules during chemical reactions using laser technology and the rapid use of a new standard of time is Alfmto seconds (10^{15} seconds). Zewail has been used pulses of laser beam of a partial vacuum in the middle of materials to study the chemistry of high- speed stages of the transition, working within the Alfmto seconds be managed after the suddenness of molecules in the interim period and then became a pioneer of so- called Alfmto chemistry using laser technology (laser Alfmto) camera and a very fast, sophisticated and very accurate to portray the ongoing chemical reaction between the molecules in three- dimensional image Alfmto time in its three dimensions, not one dimension only.

Finally, what scientist do is to identify cases of transition of chemical reactions as broken links and new links up, and the development of new chemistry carried the name Alfmto result of invention, or a new laser called laser Alfmto or laser technology and through rapid as we were filmed for the moment the chemical reaction within the atoms in the process of only one part of a billion a second, and therefore this technology and its owner, Dr. Zewail laser secrets complex world characterized by inventing something new the properties of new energy and knowledge of the movement of particles from birth or docking to know what was happening in record time is a million billionth of a second the proportion of this period to the second equivalent of one per second span of time to 22 million years ago.

To reach Dr. Ahmed Zewail of the use of laser microscopy to clarify the picture may have been the most difficult times in less than two and thanks to the time factor has been developed to see things, whether internal or external speed and one millionth of a billionth of a second.

The features of Applied Chemistry Alfmto side is represented as medical, industrial and agricultural in nature and changes in the human body, such as treatment of diseases such as cancer, diabetes, a cell can be imaged in the human body, and according to that disease can be determined in the light of the nature of these cells.

Bioinformatics

The modern scientist characterized by the advantages and new versions is in the lab trying to understand the practical environment in which he lives and solve puzzles and formulates symbolic formats for a network of information and communication for the purpose of creating and responding to inquiries, questions of science resulting from a novel link between health and genetics. The question that presents itself is to us after the introduction on the topic of bioinformatics. The term is relatively new view and bioinformatics is also new. As is the case for many of the terms introduced in modern science, scientists have so far failed to agree on a single definition of the term. It may therefore suffice here to say that the (computing life) is the process whereby the relationships between technologies, biotechnology and computer technology, for the purpose of exchanging information and experiences and the related transfer of ideas and information for the purpose of understanding of life and death of organisms and also represents the integration of mathematics, statistics, computing and life sciences for the purpose of organizing (computing life) and analysis and interpretation.

In spite of informatics is modern science that he knew very complex rooted in and accountable to the most important multi-disciplinary mathematics and computer science as well as many explanations of the science or medical studies resulting from the human genome (genetic content) for different organism as well as the use of electronic informatics to understand Genetic Network Models and the development of three-dimensional views of the complex molecules of the computer.

The basis of bioinformatics has taken different forms and the use of methods and tools for a variety of information consistent with the idea of the basic units of DNA. The gene, but at the same time vary according to the sequence of these units themselves and the simplicity or complexity.

The conditions and circumstances that prevailed in the incidence of different diseases and conditions associated with this aspect of the major change is largely in the increasing trend towards the modernization of information about human biology, including chromosomal abnormalities and related diseases for example, inherent "Down" is characterized by the fact that cell individual contains a third copy of chromosome 21, and is determine by microscopic examination, but tremendous progress has been possible to monitor changes in the DNA including the mutations responsible for many genetic diseases.

The important thing anyway is that recent data are heavily dependent on modern technologies that are designed to connect researchers, these operations become aware of the basics of bioinformatics associated with sophisticated and effective, hence the importance and seriousness of these means which undoubtedly provide the benefit greatly from this knowledge and its applications in health and genetics.

The inter-relationship between computer scientists and molecular biologists can computer world help to provide interpretation of the rules and the necessary to encrypt the DNA and therefore provide in the future and relevance of information indicated the task to determine the organism. Moreover, there is a possibility for the development of computer using DNA and the use of mathematical symbols rules when you're adding a certain amount of DNA that can store all the information in the computer world. The idea of computer life which appeared in the 1995 adaptation of DNA affords. For information processing finally, the biology is no longer limited to specialists, but it entered the computer science and scientists, who began learning these sciences and participate in the research teams of specialists to develop computer- life, as it can be envisaged that biology has become a note in terms of information called bioinformatics and the size of the smallest components of the computer-life billions of times the size of silicon chip is that characterized by very huge storage capacity and speed of processing to resolve some complex issues.

From the standpoint of the computer, the DNA and its installation of structural system is an intelligent and sober for the storage and dissemination of information and computer scientists had been accustomed to dealing with a binary digital expression of the alphabet, numbers and symbols, and diagnosed a direct letter of the alphabet of letter DNA. In order to encrypt messages, and all three relay in DNA is an information system for the manufacture of proteins and other compounds in spite of the inability to resolve the issue of structural triangular shape of proteins, despite the existence of mathematical models that serve the purpose.

Due to the fact that the storage of information in sophisticated computers to evoke a dramatically different senses, as a researcher actually lives inside a computer, a view of human life, all stored images of modern technology and not necessarily be tapped and developed. As several researchers in countries that have contributed to the success of the development of this human means and comparisons, as were studied in the genetic makeup of living non- human colon fruit flies.

The inter-relationship between scientists and molecular biologists can offer a update knowledge, interpretation of the rules and necessary to encrypt the DNA and therefore provide in the future and relevance of information is indicated task to determine the organism. Moreover, there is a possibility to develop the use of DNA and the use of mathematical symbols rules when adding a certain amount of DNA that appeared in the 1995 adaptation of DNA afford to address the NMC and finally, biology is no longer limited to specialists, but it came to the attention and scientists are beginning to learn this science and to participate in the research teams, of specialists to develop life, as it can be envisaged that biology is located in the heart of a paradigm shift led by scientists and biology became the so- called informational note.

Vital components and the size of vital billions of tunes smaller than the size of silicon chip and is characterized by very huge storage capacity and speed of processing to resolve some complex issues. From the standpoint of DNA and installation of structural system is an intelligent and sober for the storage and dissemination of information that scientists have accustomed to dealing with binary digital expression of the alphabet, numbers and symbols, and diagnosed a direct letter of the alphabet in the four DNA.

In order to encrypt messages, and all three relay in DNA is an information system for the manufacture of proteins and other compounds in spite of the inability to resolve the issue of structural triangular shape of proteins, despite the existence of mathematical models serve the purpose due to the fact that the storage of information in the accounts of sophisticated evoke a dramatically different senses, where a researcher is already in the arithmetic of human life, finds its stock, all images of modern technology and not necessarily be tapped and developed, as several researchers in countries that have contributed to the success of the Human Genome Project, the development of this means and comparisons, as were studied in the genetic makeup of living non-human colon fruit flies.

Research on the nature of genomes and the nature of the composition and organization of various scientific institutions. Acceleration in the implementation of the project, which was planned to complete within 15 years of technical progress, but shortened the time to ten years

.

Search the content of the genome to know what kind of information or material contained in the communication and diagnosis of all the genes strictly speaking, the number 100000 in the amount of DNA human as well as identifying sequences around three billion chemical bases. Study the nature of information storage in the computer and the evolution of elaborate techniques of efficient and sequence evolution that contribute to the analysis of information.

Study the effects to be set in the community and to what extent this can be achieved, and the type of response. In spite of all reported studies and research conducted by methods and techniques in various vital information as well as numerous writings in this area, there are still many other fields, and a variety of study and research. Some of these fields has not been touched so far in the country, especially the human genome projects, areas and areas to attract the attention of researchers, but it's mostly a few problems, mostly dealing with partial or subsidiary.

Diagnostic
Securing the different types of health services, preventive, curative and rehabilitation of the basic necessities of the individual and society is part of the economic and social development, the Ministry of Health before the embargo the country and the terms of reference of modern medical equipment and refurbished medical equipment, such that the health services in the country in all its aspects to the stage of qualitative and quantitative development admittedly many of the specialized agencies and international experts. That the imposition of the embargo has negatively impacted on the level of health services and spare the necessary medical supplies such as vaccines, medicines and laboratory solutions. In spite of that medicines and medical supplies but not prohibited under UN resolutions, but that the need for medicines and medical supplies that have increased due to the

deteriorating state of health, environmental and food, which led to the emergence of many diseases, and chronic diseases and malnutrition. For example, Iraq was clean and free of cholera; the disease returned and appeared again significantly in 1991.

The scarcity of essential drugs and lack of availability of the required quantities led to the deterioration of the situation of citizens suffering from chronic diseases such as sugar and heart disease, hypertension, epilepsy, kidney failure and cancer diseases.

The laboratory tests were not no better than drugs, because the lack of laboratory materials and equipment used to conduct those tests and lack of maintenance and sustaining them available because of the acute shortage of spare parts and failing to be delivered to Iraq as well as the lack of diagnostic kits necessary to conduct examinations and laboratory1 tests, all that reflected negatively on the number of tests performed annually following table shows the percentage decline in the monthly Madal to prepare laboratory tests compared to 1989.

Impediments to the implementation of the diagnostic kits (negatives)

• The lack of some raw materials necessary for the completion of diagnostic kits, including:

- Chemicals.
- Other essentials.
- Hardware.

• Difficulties in meeting the needs of the researcher:

- Chemicals.
- Services in the local market.

• Continuing attempts to obtain materials and devices from other outlets outside the country led to:

- The survival of the need for quite a few of the resources vernacular.

• The cost of scientific resources, stationery and print the necessary reports to parents of high.

- The difficulty in obtaining journals and literature of modern world.

• The area of examination and evaluation

- Assigning one for the purpose of examination and evaluation.

- Not possible to give a certificate of inspection for some diagnostic kits for the following reasons the amount of material sent for testing are limited, instabilities of some materials, lack of some modern techniques and equipment.

- Unable to implement a number of these numbers.

• A steady increase in prices of materials and equipment and the cost of sustaining an impact on services:

- Estimates of the prices offered by the researchers.

- A number of research and is now paragraph of materials and devices is the amount greater than what they have as much time of signing the contract.

• Lack of standard materials and solutions for a number of diagnostic as reference material for the purpose of comparison scanned materials and productive and the lack of number of standard delay in conducting the tests or they can not be implemented .

Chapter Four

Higher education and science

Preface

The new pattern in higher education is to ensure the quality and its management, as well as that there is quality assessment and quality evaluation and quality assurance. The term assessment will receive many meanings and connotations, while the term "performance standard" refers to the level of achievements. There is confusion between standards and criteria, while using quality assessment and quality review as synonymous to evaluation.

The terms accreditation and quality assurance in higher education differ from one country to another. The standards are used in the USA as the same meaning of criteria, but in Europe synonymous of quality assurance; whereas the quality assurance is a part of the quality management in higher education.

Various concepts of quality in higher education are used to achieve accuracy and perfection through quality management. Quality is a unique kind of performance more applicable to higher education, as well as the quality of students, that is the ability to change constantly. Another concept of quality is the ability to report the value money and this concept has become common place, especially when something fits the quality product. There are other concepts of quality, including the concession (Excellence). Quality is suitable for the purpose of fitness as well as the traditional concept of quality of higher education, has been associated with the inspection and rejection. The transformation of this traditional concept of quality in higher education to the concept of quality assurance of higher education is based primarily on the need to test the typical rates of performance and build quality management systems for higher education. With the difficulties of application it appeared extremely important for the application of total quality management in higher education and requiring participation to ensure the survival and continuity of higher education institutions.

Quality education means an estimated total characteristics and advantages of the product to meet the educational requirements of students, labor markets, society and all internal and external benefit. The achievements of quality of higher education require directions of all of human resources, systems, methods processes and infrastructure to create favorable conditions for innovation and creativity.

For quality there are several concepts such as:-

- Value for money.
- Added value.
- Transformation methods.
- Fitness of purpose.
- Consumer satisfaction.
- Enhancement
- Improvement.

There are five areas of quality of higher education:-

- A whole system of higher education.
- Foundation.
- Programs.
- Education operations.
- Outputs.

Some propose a distinction between three sets of specifications for quality:-

- Scrutiny subjects.
- Measurements.
- Inferred tools.

Austin designs two criteria for defining quality, especially in higher education, the first criteria: the view that the concept of quality in higher education, must focus on the fame and reputation of the institution or its sources, The second criteria: is believed that the definition of quality education must be enhanced and strengthened through the application of philosophy of improving quality.

The traditional concept of quality of university education is associated with the screening operations, focus on the test, then the great importance for the application of total quality management in the higher education to ensure the survival and continuity of higher education institutions.

Quality of higher education includes indicators for measuring the level of achievement in teaching and measuring of adequacy of the needs of labour markets, the number of students to each teacher, spending per students and repetition rate. It is the concept of multidimensional and dynamic levels.

Quality is important to improve the outputs of the educational process and develop the participation of the country and society, where the country has the responsibility to adopt quality standards that must show

their availability for the president of the university, dean and head of the department, may be unable to achieve on its own in certain situations.

Multiple concepts of quality as pointed by both Green and Harvey, including:

- Precision.
- Perfection.
- Constant change.

And Lim in 2001 put on mechanism for quality assurance in higher education that includes:

- Setting targets for higher education institutions.
- Compatibility with the general goals of the society.
- Test the effectiveness of quality management system to achieve the goals of higher education.

Quality depends on the context, which is called quality system and the message of the university and its goals. The quality varies in meaning depending on the:

- Components of higher education, students, university, the labour market.
- Frames of reference for quality, inputs, processes, outputs, and goals.
- Historical periods.

The quality of higher education requires the direction of all human resources, policies, systems, methods, processes and infrastructure in order to create favorable conditions for innovation of educational product.

The culture of quality and its programs lead to involvement of everyone, management scientific integrity, student and faculty member, to become a part of these programs and therefore the quality is the driving force required to push the system of university education.

The new pattern in higher education is to ensure the quality and its management, as well as that there is quality assessment and quality evaluation and quality assurance. The term assessment will receive many meanings and connotations, while the term "performance standard" refers to the level of achievements. There is confusion between standards and criteria, while using quality assessment and quality review as synonymous to evaluation.

The terms accreditation and quality assurance in higher education differ from one country to another. The standards are used in the USA as the same meaning of criteria, but in Europe synonymous of quality assurance; whereas the quality assurance is a part of the quality management in higher education.

Various concepts of quality in higher education are used to achieve accuracy and perfection through quality management. Quality is a unique kind of performance more applicable to higher education, as well as the quality of students, that is the ability to change constantly. Another concept of quality is the ability to report the value money and this concept has become common place, especially when something fits the quality product. There are other concepts of quality, including the concession (Excellence). Quality is suitable for the purpose of fitness as well as the traditional concept of quality of higher education, has been associated with the inspection and rejection. The transformation of this traditional concept of quality in higher education to the concept of quality assurance of higher education is based primarily on the need to test the typical rates of performance and build quality management systems for higher education. With the difficulties of application it appeared extremely important for the application of total quality management in higher education and requiring participation to ensure the survival and continuity of higher education institutions.

Quality education means an estimated total characteristics and advantages of the product to meet the educational requirements of students, labor markets, society and all internal and external benefit. The achievements of quality of higher education require directions of all of human resources, systems, methods processes and infrastructure to create favorable conditions for innovation and creativity.

For quality there are several concepts such as:-

- Value for money.
- Added value.
- Transformation methods.
- Fitness of purpose.
- Consumer satisfaction.
- Enhancement
- Improvement

There are five areas of quality of higher education:-

- A whole system of higher education.
- Foundation.
- Programs.
- Education operations.

176

- Outputs.

Some propose a distinction between three sets of specifications for quality:-

- Scrutiny subjects.
- Measurements.
- Inferred tools.

Austin designs two criteria for defining quality, especially in higher education, <u>the first criteria</u>: the view that the concept of quality in higher education, must focus on the fame and reputation of the institution or its sources, <u>The second criteria</u>: is believed that the definition of quality education must be enhanced and strengthened through the application of philosophy of improving quality.

The traditional concept of quality of university education is associated with the screening operations, focus on the test, then the great importance for the application of total quality management in the higher education to ensure the survival and continuity of higher education institutions.

Quality of higher education includes indicators for measuring the level of achievement in teaching and measuring of adequacy of the needs of labour markets, the number of students to each teacher, spending per students and repetition rate. It is the concept of multidimensional and dynamic levels.

Quality is important to improve the outputs of the educational process and develop the participation of the country and society, where the country has the responsibility to adopt quality standards that must show their availability for the president of the university, dean and head of the department, may be unable to achieve on its own in certain situations.

Multiple concepts of quality as pointed by both Green and Harvey, including:

- Precision.

- Perfection.
- Constant change.

And Lim in 2001 put on mechanism for quality assurance in higher education that includes:

- Setting targets for higher education institutions.
- Compatibility with the general goals of the society.
- Test the effectiveness of quality management system to achieve the goals of higher education.

Quality depends on the context, which is called quality system and the message of the university and its goals. The quality varies in meaning depending on the:

- Components of higher education, students, university, the labour market.
  - Frames of reference for quality, inputs, processes, outputs, and goals.
  - Historical periods.

The quality of higher education requires the direction of all human resources, policies, systems, methods, processes and infrastructure in order to create favorable conditions for innovation of educational product.

The culture of quality and its programs lead to involvement of everyone, management scientific integrity, student and faculty member, to become a part of these programs and therefore the quality is the driving force required to push the system of university education.

Total quality is the tool and process for practical application, which aims at achieving a culture of continues improvement in order to appease and please consumers and costumers, have many perceptions and concepts, including:-

- Philosophy concept.
- Tools and processes.

- Continuous improvements.
- All employees.
- Gratification, appease and please of customers.

Dealing with the comprehensive quality of whole educational system requires encouragement of individuals to participate in decision making.

Multiple terminologies are used for classification of countries according to economic conditions, social, educational, cultural. The backward demonstrates the lack of hope in economic reform in the developing countries, while in the under-developed countries it refers to the poverty of performance. [The third world means the (non-developed countries), the first world means (the industrialized countries of Europe and America), and the world II means (former socialist countries)].

The use of the entrance to the execution of quality assurance in higher education is an effective if certain requirements are available, such as qualified faculty members that devote themselves to work all the time at the university, available administrative services, and university leaders realize the importance and necessity of quality assurance.

The quality assurance is a process which focuses on quality measurement procedure for the institution or program and refers to a range of activities, methods, and procedures. The term is used alternately with quality management, quality assurance and quality control aimed at controlling the process and removing the causes of unacceptable performance. The overall quality assurance indicates:

- Policies and directed processes.
- Description of all the systems sources, and information of higher education.
- Planned and organized reconsideration of the institution or programme.
- Planned and organized activities that apply the quality.

- Evaluating the ongoing process (assess, control, security, maintenance, improving the quality of education system).

The principle of quality assurance in the higher education contains general polices and processes geared towards providing all help to achieve quality and preservation, they are as follows:

- The meaning of quality in higher education.
- Entrance performance to ensure quality.
- Quality inspection system.
- Organization of education.
- Management plans.

Quality management in higher education requires a system in several sequential steps that are serialized as follows:-

- The task of identifying the goal of the university.
- Determining the functions of the university.
- Defining the objectives of each post.
- Establishing a management system and quality assurance.
- Placing a system of quality inspection.

Defining the mission of the university must be conducted through cooperation and understanding the vast majority of the university community in accordance with laws and aspiration of state, the university of modern concepts and futurology realism of the university. The functions of the university are supposed to be modern (teaching, scientific research, community service) with clear objectives for each these posts. Teaching has goals, such as curriculum, innovative and effective with modern methods of teaching and distinguished health teaching climate. The scientific research has objectives that deal with increase of the productivity of research, publishing with marketing and modern skills. The goals of community service function are represented by accommodation of the needs of society and cooperation with other organizations of the society.

The recent trends in quality management are working to avoid a narrow view and work on measuring the output of university education according to different tools that provide educational services at the level of bachelor.

Developments of integrated system for the university and institutions management plans, and the application of the structure of modern management plans for teaching and scientific research and community service, within the clear objectives and specific strategies on several levels. These parameters require financial resources and follow, sequence in the preparation of plans at different levels of the university.

Quality management system requires several consecutive steps, including:

- Identification of the message of the university, college, department.
- Determining the functions of the university, college and department and their relative importance.   237
- Defining the objectives of each function of the university or college or department.
- Identification of the system of quality management.
- Preparation of quality inspection system (Evaluation of the performance of the university, college, department).

The conditions for the quality assurance system include the following:

- Eligibility of faculty members with appropriate degree.
- Providing administrative services and good electronic tools.
- Capacity of university current factors in the quality of higher education.
- Using and adopting of scientific promotion according to academic potentials.

The inspection and screening process of quality carried out according to evaluation performance of the university and depend on the quality of the new file that contains:-

- Message of the university.
- Specific objectives of teaching and research.
- Community services.
- Quality management system.
- Performance standards.

The evaluation process is necessary as an integral part of the process of developing and colleges affiliated with. It is also necessary to identify and to achieve the objectives of those the tools that are used to measure the advancement of the university could be summarized according to the followings:-

- Measurement of the efficiency of output (graduates).
- Scientific level.
- Scientific achievements.
- Research.

Whereas, the mechanisms used for evaluations depends on description and governance. Description is carried out by gathering information, whereas governance is done through combining certain characteristics that are needed for the evaluation, in addition to the identification of certain principles to be implemented, such as the purpose of the evaluation and selection tools for evaluation.

The evaluation of higher education is an integral part of the educational process. Thus, it becomes an essential requirement of all international academic accreditation bodies. The evaluation of educational output was one of the conditions for academic accreditation according to the "Abet" document. The importance of the evaluation encourages 79% of the universities and institutions in USA to use this parameter.

Ranking criteria and weights depend on several indicators, such as:

- Academic achievement.
- Nobel prizes.
- Fields which obtained the awards.
- Researchers who have been accredited.
- Articles that were published in the journals, nature and science.
- Articles that were published.

These decisions have been taken by the world body to evaluate world universities, and then adopted the following criteria:

- Production of scientific research in various scientific fields.
- Quality of graduates from universities and their scientific levels.
- The facilities provided by universities in their areas of work.
- Contributions made by the universities of modern knowledge.
- The use of science and technology.
- Universities ability to use information and technology.
- Information posted on the university.
- The performance of universities.

Total quality is the tool and process for practical application, which aims at achieving a culture of continues improvement in order to appease and please consumers and costumers, have many perceptions and concepts, including:-

- Philosophy concept.
- Tools and processes.
- Continuous improvements.
- All employees.
- Gratification, appease and please of customers.

Dealing with the comprehensive quality of whole educational system requires encouragement of individuals to participate in decision making.

Multiple terminologies are used for classification of countries according to economic conditions, social, educational, cultural. The backward demonstrates the lack of hope in economic reform in the developing countries, while in the under-developed countries it refers to the poverty of performance. [The third world means the (non-developed countries), the first world means (the industrialized countries of Europe and America), and the world II means (former socialist countries)].

The use of the entrance to the execution of quality assurance in higher education is an effective if certain requirements are available, such as qualified faculty members that devote themselves to work all the time at the university, available administrative services, and university leaders realize the importance and necessity of quality assurance.

The quality assurance is a process which focuses on quality measurement procedure for the institution or program and refers to a range of activities, methods, and procedures. The term is used alternately with quality management, quality assurance and quality control aimed at controlling the process and removing the causes of unacceptable performance. The overall quality assurance indicates:

- Policies and directed processes.
- Description of all the systems sources, and information of higher education.
- Planned and organized reconsideration of the institution or programme.
- Planned and organized activities that apply the quality.
- Evaluating the ongoing process (assess, control, security, maintenance, improving the quality of education system).

The principle of quality assurance in the higher education contains general polices and processes geared towards providing all help to achieve quality and preservation, they are as follows:

- The meaning of quality in higher education.
- Entrance performance to ensure quality.
- Quality inspection system.
- Organization of education.
- Management plans.

Quality management in higher education requires   a system in several sequential steps that are serialized as follows:-

- The task of identifying the goal of the university.
- Determining the functions of the university.
- Defining the objectives of each post.
- Establishing a management system and quality assurance.
- Placing a system of quality inspection.

Defining the mission of the university must be conducted through cooperation and understanding the vast majority of the university community in accordance with laws and aspiration of state, the university of modern concepts and futurology realism of the university. The functions of the university are supposed to be modern (teaching, scientific research, community service) with clear objectives for each these posts. Teaching has goals, such as curriculum, innovative and effective with modern methods of teaching and distinguished health teaching climate. The scientific research has objectives that deal with increase of the productivity of research, publishing with marketing and modern skills. The goals of community service function are represented by accommodation of the needs of society and cooperation with other organizations of the society.

The recent trends in quality management are working to avoid a narrow view and work on measuring the output of university education according to different tools that provide educational services at the level of bachelor.

Developments of integrated system for the university and institutions management plans, and the application of the structure of modern management plans for teaching and scientific research and community service, within the clear objectives and specific strategies on several levels. These parameters require financial resources and follow, sequence in the preparation of plans at different levels of the university.

Quality management system requires several consecutive steps, including:

- Identification of the message of the university, college, department.
- Determining the functions of the university, college and department and their relative importance.
- Defining the objectives of each function of the university or college or department.
- Identification of the system of quality management.
- Preparation of quality inspection system (Evaluation of the performance of the university, college, department).

The conditions for the quality assurance system include the following:

- Eligibility of faculty members with appropriate degree.
- Providing administrative services and good electronic tools.
- Capacity of university current factors in the quality of higher education.
- Using and adopting of scientific promotion according to academic potentials.

The inspection and screening process of quality carried out according to evaluation performance of the university and depend on the quality of the new file that contains:-

- Message of the university.
- Specific objectives of teaching and research.
- Community services.

- Quality management system.
- Performance standards.

Brain Drain

Migration of scientists is due to various reasons whose effects are overlapping sometimes; and these create psychological climate related to scientific and incentives conditions. Some of the reasons are social in the homelands, others are materials; may be related to the needs of living.

The phenomenon of brain drain has motives of social, political, and personal nature. The social is characterized by difficulties that faced the developing countries in strengthening the shaken scientific planning in developed countries. One way to keep scientists from migration is to treat the fundamental faults by working to link with national policy, to introduce the idea of scientific planning, providing possibilities for scientific work and atmosphere.

It is an unfortunate fact that the money spent on scientific research and development of the university, in all Arab counties up to 260 million dollars only, while the states of Western Europe during the sixties spent 6 billion dollars per year. United States of America spent 24 billion dollars during the same period. The value of spending on research has increased with the beginning of the eighties to nearly 40 billion dollars. This has encouraged migration of scientists, for example, Iraqis abroad had been attracted to the atmosphere of academic and scientific facilities, methods and the possibility of attending scientific conferences, symposia, magnitude of printed and published   by   specialized   magazines   and   periodicals.
The phenomenon of brain drain is the most important global problems which recorded at the international level and regional level as stated. Recent studies indicated that the organization for economic cooperation and

development which includes 30 industrialized states, that the immigrant enjoyed   a degree of education.

It is worth mentioning that the phrase "Brain Drain" derived by British was used to describe the loss of scientist, engineers and doctors. The UNESCO defines immigration as kind of abnormal types of scientific exchange between the states, as a reverse transfer of technology. The Gulf center for strategic studies in 2004 indicated that western counties had attracted to the west no fewer than 31% of the brain drain for developing countries by about 50% of doctors and 32% of the engineers, and others within the intellectual trends. The west is perceived to the issue of brain drain from standpoint that they reproduce underdevelopment in developing nations.

The risk of brain drain may vary from one state to another, but effects remain similar - in that (brain depriving in human resources).

The "World Organization for Migration" estimates those developing countries supporting the United States, Europe and South Asia at 500 million dollars annually. On the other hand, the World Bank estimates that one hundred thousand foreigners from industrial nations are working in Africa at a cost of four billion dollars annually. Certain social values prevailed in the traditional farming communities would also decline; especially that migration was characterized as external migration of males. Therefore, the migration process may lead to partial destruction of the wealth of mankind.

Brain drain that began after World War II included developing and developed countries, spearheaded by Britain, France, Germany, Sweden, Switzerland and Japan. The United States of America was not included, but limited to become   a terminal brain drain, from other countries. From a historical stand point, this kind of migration was due to Phoenician and Golden ages of Greek, Roman and Arab civilizations.

Theoretically, migration and mobility of scientists across the centuries from country to country, is considered as one of the features of scientific development. The UNESCO in 1955 considered brain drain as "the impact of migration from the effects of human solidarity".

It is important to study the effects of brain drain on the Arab countries in the perspective of a strategy to develop higher education. The competencies move highly qualified group of individuals from one Arab country to another. This brain drain has not received international attention until only at the end of the sixties and seventies. It came in the form of transmission of certain competencies to the industrialized countries, where brains moved abroad to more advanced societies to increase productivity; but at the same time it caused a loss for the country of origin.

It is estimated that Egypt had provided about 60% of immigrates to USA, and Iraq has increased its share significantly after the nineties, followed by Syria, Jordon, Palestine. The UNESCO has chosen Egypt from among those most affected by the brain drain, but did not contribute dramatically to solve this problem.

Egypt has used students returning from study abroad in the appropriate places, but Iraq was the first to issue legislation to participate in solving this problem

Arab brain drain

The Arab brain drain was launched as a phenomenon specifically since the nineteenth century and especially in Syria, Lebanon and Algeria. At the beginning of the twentieth century the immigration increased, especially during the First and the Second World Wars. Migration of doctors, engineers and Arabs scientists to Western Europe and the United States, arrived in 1976 to about 24000 doctors, 17000 engineers, and 75000 researchers. In the last fifty years, the migration from the Arab countries counted between approximately 25 to 50% of the total of Arab drain to the United States of America, Canada, and Britain - the two countries that attract more than 75% of the Arab immigrants. Moreover, 50% of doctors and 23% of engineers and 15% of researchers from of the total Arab competencies who graduated in the last fifty years now migrated to Europe, America, Canada and Britain. Only 54% of Arab students studying abroad returned to their countries. The

Arab doctors working in Britain constitute about 34% of the total number of doctors working there.

The losses reached the Arab states as a result of the Arab brain drain to about 200 billion dollars annually, according to the report of the Arab Labor Organization in 2006. The Egyptian brain drain is representing a big problem after the figures, announced by the large body of statistics, have reached 824 thousand immigrants. From the competencies of Migration to Europe and the United States, Canada, Australia, the scientists represent more than 10,000 immigrants, and the rest are in the areas of medicine, engineering, basic sciences, agriculture and humanitarian sciences.

The report was cautioned by the Arab League in 2001 that Arab countries lost two hundred billion dollars as a result of scientific brain drain. The report pointed out that Western countries are the biggest beneficiary of more than 45 thousand Arab holders of certificates and qualifications.

The working paper submitted by the Department of Immigration and Population Policy in the Arab League for the first meeting of ministers was that the Arab migration has positive and negative results. The, countries of the Organization of Economic Cooperation for Development hosts million immigrants from Arab with advanced degrees, while five thousand doctors migrate annually to Europe.

The Arab Labor Organization issued a report in which it indicated that scientists are half of the immigrants to western countries, and the study indicates the reasons, such as the tyranny of political and governmental interference in the affairs of the universities of Arab States, and the lack of freedom of scientific research, underlying motivation for Migration. Also according to the study of the Gulf Centre for Strategic Studies issued in 2004, the western states capitals attracted no fewer than 450,000 of the Arab minds. The Arab countries contribute to 31% of brain drain from the developing countries to the west of capital by about 50% of doctors and 23% engineers and 5% of scientists. The memo distributed by Arabic Labor Organization, indicated that the past ten years have witnessed an increase in the number of immigrants of Arab scientists, especially in the fields of medicine and engineering and science, to work in Europe. More than 2600

Egyptians working in scientific positions in countries such as America which came in the first place, while Germany came in the second and Canada in the third and Spain in fourth and France in the fifth. It is clear that Egypt is the biggest loser of the brain drain in various professions.

At a time when one study of the United Nations Development Program showed that in the period (1998-2001) more than 15,000 Arab doctors emigrated from their countries. According to statistics issued by the United Nations Organization that almost 50% of doctors and 33% of engineers, 15% of the total drain of scientists from the Arab graduate, 70% of migrants attracting the United States, Canada and Britain. The Iraqi scientific inefficiency leakage and migration at these rates reached maximum levels after the events of (1991) and became a higher education problem. In that era the development led to the departure of scientific goals followed after the migration of large scientific events in (2003) and especially in the year (2006)Brain drain can be summarized in general conditions: they are economic, social, political, cultural, intellectual, psychological and scientific, including:

• Low-income and poor living material for skilled people.
• The lack of balance of the educational system.
• Loss of the link between education systems and development projects.
• Lack of appreciation of the distinguished scientists.
• The absence of adequate scientific research climate.
• Restrictions and barriers on studies, ideas and research.
• Granting of skilled people in wages and financial incentives.
• Encouraging and facilitating of assistance to ensure the provision of adequate housing and services needed.
• Cooperation of international and regional organizations such as UNESCO to set up projects and academic centers to help brain drain.

The Arab countries, measures to curb immigrations and in particular

Iraq and Egypt by legislation and measurement to reduce the brain drain, including the re-adoption of Law No. 189 of competencies of 1975. But the calamities that afflicted Iraq during eighties and nineties and after 2003 led to the leave of thousands of Iraqis from the country, most of them are skilled and scientific qualifications.

The Arab Parliamentary Union at its tenth conference in Khartoum on February 11, 2002, identified a number of reasons for the emergence of brain drain and the ability to absorb skilled people, including:

- The poor economic returns of competencies.
- The lack of balance in the educational system.
- Political and social instability.

The economic factors that lead to the phenomenon of brain drain are:
- Lack of physical return of workers from skilled people.
- Failure to provide material resources.
- Diminished access to scientific inefficiency.

However, the methods of dealing with the file of brain drain in general led to provoke the drain itself. Some countries have poured extensive privileges to these competencies, wishing to return to serve their country and possibly met with initial success of these invitations, but did not continue in its competence. To the opposite side, China clearly considered a lot of qualified scientists worldwide, in the framework of a political project, considered to build the nation by:

- Academic and scientific freedoms.
- Giving scientific bodies and academic freedom of expression.
- Providing the necessary potentials to reach various items of scientific knowledge.
- Separating education from politics.

Iraqi universities are able in very limited ways to build information networks that can be developed to provide Internet services to workers and students. The ratio of the number of students to the number of computer available is still very high and where there is no general frame work for the university plans in information technology is noted. A state of chaos exists in the process of developing these plans which are often duplicated. In the light of randomness must be a general framework that defines the requirements and competencies to be developed.

Advanced Education

In (1964), 119 countries approved at the general conference of UNESCO, the recommendations that include the different forms of education outlying school and adult learning must be considered an interal part in the education system, the opportunity for males and females to continue to lifelong education. Patterns of continuing education in Iraq universities remain a valuable and suffer in Breach of the continuous weakness in performance, despite the long period exercised.

Communications equipment and satellite

Communications satellites are able to grant broad prospects for the educational process at all stages of education, especially, in higher education. The potential has become possible to lecture from the university to house or transferred to another university.

University of Hawaii Islands began testing the use of satellite ATS- I to transfer voice message and printing between its various islands in the year (1971). Then established the university plans to send ground - receiving television broadcasts in educational exchanges between libraries and medical conferences, student and teacher training joint research. In (1971) satellite ATS-I was used to provide medical treatment and educational programs for rural schools and some guidance to some medical lectures in the college of Medicine. University of Washington has also further tests at Stanford University in conjunction with Brazil using the moon ATS-6, as well as experience of the territory Rocky Ponte Moon using the ATS-6 and the experience of using the territory Appalachian Satellite.

In Ottawa, Canada, Stanford and California Universities in the United States, exchanges of experiences between teachers via satellite and mutual distances are often performed.

The experience of the Arab satellite ARBSAT from the Arab Satellite Communications Organization will be   a successful solution to

the problems of education in rural areas; also it will help to compensate and cope up with the serious shortfall in the preparation. There is no education of this kind in Iraq.

Data Banks are information outlet to the problem of traditional storage and retrieval of information which can be stored in these centers, using one of the elements described as document number, copyright, title research, objective data banks to link centers and scientific institutions and universities. These are tools of opening the doors in front of knowledge and information gathered by international experiences. In Iraq there are no techniques and techniques of terminology to this type of education.

.

The developments were strengthened in technology and communications to a pattern of Open University education, because of its reliance on techniques of knowledge and information, given the rapid development of societies and the transition to a knowledge society. Knowledge society has expanded greatly in introducing this type of university education. The justifications mentioned previously include:
- Integrated use of other media technology.
- Defeat of the many obstacles to the normal university education.
- Submission services to individuals of all ages.
- Defeating of the barrier location.
- Allowing the teacher to work and learn.
- Entrenches a culture of continuous learning.
- Development of opportunities for developing the performance of workers in state institutions.
- Increasing social demand for tertiary education.

There is a clear correlation between knowledge society and technological knowledge, are two sides of one coin. The role of the knowledge society is clear and consists of dimensions as a result of spectacular progress in technology applications, where communications .

.

Turning to technology revolution is one of the important creations performed by rights in the late twentieth century and the beginnings of the twenty-first century, then this session has widened in areas essential to humans, including:

- chemistry, medicine
- Bio-Engineering
- Other sciences

Substantial changes in the pattern of human life become a modern mind and thought that is the basis of profit and invest therefore there is sufficient justification, emphasizes justification, importance and necessity of Iraqi universities to take this kind of education:

- Rapid technological evolution.
- Large increase in the numbers of educated and willing to education.
- Democratic education and rights of citizens to education.

To meet the challenges of community learning, it is required to work with attention to information technology, knowledge-building rules, modern communications networks, integration of technology in teaching and learning processes, and research. The success of these universities and transforming them into learning societies, depend on the extent of interest in the development professionally, and adoption of the principles of participation planning and stimulation of the use of modern technology and transforming classrooms into environments for active and effective learning.

The challenges of the knowledge society depends on the nature of the nature of the information and knowledge which will be published daily in different parts of the world and very quickly, knowledge of various types, including the globalization of knowledge and virtual knowledge and technology knowledge.

Iraqi universities and scientific research

Scientific research in universities is inseparable about the problems of Iraqi society, where research aims to obtain a degree or promotion, its role in the production of knowledge and solving problems is very limited, and worth noting that universities operate on the developed world.

- Scientific research and its role in the production of knowledge and solving problems.
  - Development a clear policy for scientific research.
  - Availability of information to help scientific research.
  - Service sectors of society productivity through scientific research.

The graduate students in Iraqi universities are an important part of the manpower involved in scientific research, those students shape the future researchers, therefore these universities, create conditions of physical, moral, appropriate to attract. Despite the expansion of universities and increasing umbers of graduate students, the development of scientific research is still limited and attending school is in difficult circumstances.

Human manpower related to scientific research in Iraqi universities is characterized by a strong current reality with the following:

- The opportunity for holders of doctorates from the newly graduated in training and direct involvement in training is very limited.
  - Scarcity of complementary research teams.
  - The proportion of the number of students to the teachers of acknowledged international proportions is high.
  - Preoccupation with a large number of teaching staff to work overtime.
  - Opportunities for research assistants and technicians for training in developed countries are very limited.

Available statistics indicate that Iraq spent in 2003 on research and development rate of 0.4 % of GNP, and the universities have contributed hitting 31% of the volume of expenditure on research and development.

The analytical study on the budgets of university official in 2003, indicates that the expenses of scientific research (to support research projects, publications, journal and scientific conferences, books and periodicals) had reached in universities, compared with expenses of scientific research and development, in the year (2000), increased by 1.4% as it in (2000) and that most the increases were concentrated in the creators of scientific research (supporting , research projects, conferences and publishing missions and training).

In Iraq, the scientific journals are accused of containing low scientific-level articles. These journals adopt a committee to evaluate essays and articles, publish sometimes low quality or of questionable value and validity, and lack of scientific mentality as well the tradition of pluralism.

A number of studies approved in developing countries, which can be applied in Iraq, particularly those that rely on international scientific cite, is not convincing in assessing the scientific production of the third world, including Iraq. The measurement and analysis of the overall scientific production of these states scientifically is impossible, since no rules are approved to the local scientific work. Most of those who tried to estimate the scientific production and dissemination of scientific information are dependent on the rules of international (Institute of Scientific Information IST) and the adoption of the Guide to international scientific cite SCI.

Statistics indicate that significant portions of tie work of researchers published locally are distributed rarely outside national borders. Even if the deployment of Iraq's scientists took place in the research journals of international scientific reputation, these articles would not be quoted as articles of Western scientists and col1gues, thus it is clear that this phenomenon is different and arguably broad. Other statistics indicate that more scientific production in Iraq in the field of pure science researcher at an article published annually is mostly inside Iraq.

The dissemination of scientific output in local reality represents a strategic decision by the researcher. Local publishing is easier and more secure than messaging to a foreign scientific journal for publication

therein. It also allows the researcher to identify himself to readers of local colleagues, students and others who would never have to familiarize themselves with the research if published in   a foreign magazine.

The articles published locally, read and quoted faster than those published abroad and thus 60% of local authorities cited a recent research, while the proportion of foreign modern references was declining.

Statistics of the earlier Ministry of Information show a total of that more than 44 scientific evaluated journal had granted the right to publish and issued by:

- The earlier ministry of Information.
- The Ministry of culture.
- The Ministry of Higher Education and Scientific Research and its institutions.
- Associations and trade unions.
- Other ministries.

These journals have been granted the right to survive in Iraq. Most of them have been given to the associations, trade unions and official institutions and semi-official activity in the various aspects of scientific and technological advances.

This large number of journals has not succeeded in filling the void created by the need, not quenched thirst of researchers who are looking forward to science and knowledge and the reasons to go back to:

- The Ministry of Higher Education and Scientific Research issued- 32: some annual journals and others quarterly on semi- annual basis.
- Some journals published sporadically in the dates which are not fixed.
- Some journals are issued without serious preparation and real efficiencies.

- Thirty two specialized scientific journals that are published by the Ministry of Higher Education and Scientific Research were not chosen clearly; clear follow- up has not been issued.

- Low percentage of publishing attendance and continuation, is finding their way with difficulty and hardship to the various impediments and obstacles.

Besides, the journals or periodicals of the Ministry of Higher Education and Scientific Research, that issued according to decision No. 4660 of 5/7/1998, by council of ministers, 44 periodicals magazine called the evaluated , issued by colleges and universities, associations and trade unions, federations and scientific institutions scientific centers and some of the specialized services.

Evaluated journals are characterized and listed as follows:

- Mostly depend on the fields of specialized fields such as medical, legal, pure science, engineering and pharmaceutics, electronics, agricultural and veterinary

- Some of these journals rely on strict specialization fields such as endemic diseases and digestive system, chemistry, social science and embryos research and infertility treatment.

- Some of the valuated journals are dependent on areas of scientific and humanitarian, especially those issued by private colleges and some trade unions and centers.

- There is no documented information about continuation of these magazines and continuity and on the nature of issuance (annual, quarterly, semi- annual).

- These journals do not cover all specialized fields; lack of the life of sciences, biotechnology, genetic engineering, computer engineering, architecture and others.

- The following bodies publish specialized journals of scientific research inside Iraq and in many specialized fields
  - The Ministry of Higher Education and Scientific Research 32

- Scientific societies, colleges, universities and Arabic scientific.

• These journals suffer from:

- Financial difficulties.

- Limited distribution

- Non-programmed continuations.

• Absence of uniform methodology for the titles

• There are no uniform regulations for publication.

• There is no clear formula for the dissemination of scientific research and documentation of a confidential nature (limited circulation) well as the methodology documented.

• There is no danger in the dissemination of scientific research inside and outside   Iraq. Responsibility for issuing magazines published by the Ministry of Higher Education is transferred to associations and trade unions which have long experience in the field of scientific publishing.

In the light of what has been referred to, could raise the following proposals and recommendations:

• Support for scientific journals:

- Increase the number of scientific periodicals published in Iraq.

- Increase budgetary allocations to cover the cost of issuance and the development of permanent secretaries.

- Linking the information network with international and locate them.

- Adoption of the principle of part-time to oversee the journals.

• Relationship of journals outside Iraq.

- Urging scientific journals in the country to adopt a more open policy towards foreign searchers and invitation from abroad to contribute to the publication.

- Facilitating access to the local science publication by external scientific societies.

- Coordination with international data banks for more attenuation to science product locally.

- Allowing the local journals and researchers to have faster access and easier.

- Journals and international information network with respect to international information network requires:

- Expanding and developing the use of the network and move away from the methodology adopted at the present time.

- Allowing professors, researchers and graduate students access to the facilities of the network.

- Allowing professors and researchers access to their sites in the network.

- Reconsidering controls withholding some of the sites of foreign scientific beneficiaries inside Iraq.

- Design of local journals

Central issue instructions to deal with the magazines and periodicals could be organized as follows:

- Unifying titles of Iraqi journals (the official journals) issued by the Ministry of Higher Education, approved the methodology according to the following passages:

- Remember first the magazine.

- Remember second the word Iraq

The reform of university education in Iraq

The problem of reform of university education universally is an old one of this type of education. Now it became the determination problem in the world, especially in the Third Millennium, where grappling the strong social, economics powers. Education remains in this race can not even get out of it, contributes to find a balance between them as needed without a doubt to face non- traditionally and radically.

Higher education provides the society, knowledge and skill; consolidate the values and thus behavior in other words. Higher education contributes to development of physical and value of man and society and then determine the levels of civilization is thus a tool for change with moral advantages and civilized way.

The present stage, which represents internal and external challenges, imposed a reform of higher education stemmed from an objective and realistic understanding of the problems of this type of education, according to data accurately diagnosed as evidence of action.

The Universities of higher education are one of the key ingredients of modern states, so each state according to their levels of development and the development of higher education, owing to the fact that this professional education, leadership, includes:

- Availability of the possibilities and equipment needed for higher education.
- Identification of new and modern targets for higher education.

Moreover, the reform of university education in Iraq must be compatible with the social transformation that is happening in Iraq. The university can play a role in qualitative production, acquisition, resettlement and dissemination of Knowledge, contributing to the development process in all dimensions of economic, social and cultural right.

In addition, the Iraqi universities to face new challenges and overcome the traditional tasks of teaching, research and public service and to exercise their duties to develop themselves to social service entity, to achieve social goals because they are at the top of the educational ladder. The future of Iraq in the near term and long tern depends on higher education as a way to prepare specilized manpower and generate thought and preparation of researchers and leaders. The issue is very clear that Iraqi university reform is required according to clearly defined strategies and reports on human development and Arab world.

Iraqi university reform system understands the quality of higher education affected by factors and exciting clinnels, including

- Lack of clarity of vision and goals.
- The absence of clear policies governing the educational contexts.
- Independent universities turned into an arena for political and ideological conflicts.

- Promotion certain political currents by the ruling power.
- Department of universities is ruling by political logic.
- Open-door policy in the acceptance of students.
- Focus on the side of quantity rather than quality.
- Reduced spending on higher education.

Thus real reform of this system requires:

- Free university education institutions from shed of the Iraqi government and will bear the responsibility in the development of this education.
- Reforming the structure of higher education to build an Iraqi characterized by diversity and flexibility.
- Deployment of higher education regardless of their spending.

On the other hand, reform of university education is supposed to address the foundations of this education by:

- University teachers.
- University student.
- University infrastructure.
- Elements of university education
  - Acceptance of university student
  - University curriculum
  - Graduate studies
  - University research

This major reform was built up with the requirements with the requirements at the institutional and personal level by fixed time plan, five or ten years. The first step might be to overcome obstacles, the second for starting the advancement of education versus constraints, the shortage of quantitative and qualitative deficiencies of teachers, shortage of quantitative and qualitative scientific journals, laboratory equipment and determination the optimal for students in each university, to develop and expand new studies in advanced scientific disciplines, filling the acute shortage of Staff, continuation missions and leave fellowships, to determine the percentage of student to teachers to nearly the accepted ratios in the world, and hold universities theory teaching who have the title of "teacher" and over.

One of the academic problems awaiting solution is a study of the optimal size of the Iraqi universities and to address the problems and obstacles faced by the University of Baghdad. It is now necessary to divide the University of Baghdad, depending on the size, specialization and location.

The first Conference of Higher Education and Scientific Research in 1972 was carried out to respect higher education and scientific research, followed by the deepening of the reform of higher education symposium in 1981. The result of the symposium was the appointment of new minister who promised to change the decentralization of higher education and return to the annual educational system and the abolishment of the course unit system. In 1989 the Ministry of higher Education presented a paper on the reform of higher education and Scientific research, pointed out that the University of Baghdad characterized by complex problems in higher education such as the low scientific level of its graduates and then prepared a plan reform higher education in 1994 focused on expansion in graduate studies to compensate foe the missions, then the conference of 2004 was held but is not done as hoped.

University Reform and its Programs

The UNESCO has chosen seven cases involving university reform:
- Evaluation of the needs.
- Professional unemployment.
- Lack of jobs.
- The lack gap in financial resources.
- Imbalance in the equalization of educational opportunities including the limited participation of women.

- Geographical discrimination and accommodation of poor students.
- Difficulties of the university administration.
- International cooperation.
- Economic and social factors.

The contribution of higher education to the development in the Arab countries depends on an effective higher education that requires reform.

The report issued in 2000 by "Working Group" proposed urgently to increase the quality of higher education in developing countries. The president of the World Bank, James Wolfensohn described the report as road map for the development of policy makers.

The process of evaluating the efficiency of the Iraqi universities

- In 1992, the Ministry of Higher Education developed a system of quality control for performance, measurement and evaluation of university. The file has been building of the performance based on the Iraqi experiences in the field of measurement and evaluation.

- The year 1992-1993 was the first year of the application of evaluation file until 1995-1996. On 20/9/1997 the matter was referred to the committee formed to evaluate the performance of the office of financial supervision.

- In 4/12/1999 the ministry has formed a committee specialized in measurement and evaluation to reconsider the evaluation file in the light of the specific objectives of the Ministry and the office of financial supervision and the universities and the specific standards determined globally.

- It turned out that the commission in charge of the evaluation the file has faced difficulties for universities to apply, since:

- The file is containing a large number of elements for evaluation,
-

which amounted to more than which led to confuse the university in providing information and accurate specific data scientifically

within the time limit   The file focused on
the input more than on output and other processes.

• The file did not take into consideration the specialization of each university especially the modern ones.

• The aim of the evaluation file was to create a great fair competition between universities, leading universities to move from the credibility and objectivity in data entry, some of whom have been exaggerated by the data submitted.

• The statistical equations were unclear and not understood by representative of universities.

• Based on the above points, the file was reviewed and corrected to specialized three files (file at the university level, file on the college level, file of scientific department level), and taking into account:

- Evaluating of the efficiency of the system: (inputs and outputs and other processes, or so- called areas of the system).

- Evaluation the building system: (organizational structure, university services, teaching, students, curricula, teaching methods, scientific research, social services). The file contains 61 universities elements of the educational system, and 98 for colleges, 97 for scientific departments. These elements have been quantified, monitored and then compared the estimations. It took in preparation, construction and application of the evaluation files and building databases, auditing, analysis and writing more than two years. But this effort had vanished during the war and ended in 2003 due to the loss of files relating to this subject.

• In 2004 a central committee was formed in the ministry to re-evaluate a scientific supervision and evaluation files, then was reformulated in the light of the following criteria:

- Global standards for total quality, particularly the American and British universities.

- Standards set by the Union of Arab Universities.

- The criteria used by some universities, especially the Arab (Egyptian universities and Gulf universities).

-        The goals of education policy in higher education and scientific research

-            The goals of the Iraqi universities.

• The evaluation files have been updated and included current areas and the following variables:
  - Goals
  - Infrastructure and buildings
  - The financial aspect
  - The administrative and organizational structure
  - Library and information sources
  - Scientific research
  - Community service
  - Student activities and services
  - Faculty members
  - Students
  - Programmes and academic disciplines

The number of elements of evaluations is 95 for universities, 113 for colleges, and 81 for scientific departments.

• The evaluation files filled for the academic year   2004-2005 by universities, colleges and departments of scientific authority.

After the completion of the database, the analysis was carried out and then the reports written for each unit of the university and college, department, and comparison of their performance levels according to specific criteria was carried out to identify the sequence with their rank.

• Establishing a quality assurance center at the ministry, fully independent from the universities.

• The creation of separate departments for quality at the level of colleges and universities.

• Provide financial and efficient human resource materials, and techniques necessary for work.

A proposal to improve the quality of higher education in Iraq

The new trends in the measurement of quality management and quality components to provide educational services, which can be used and adopted on the following criteria:

Table

Quality standards of higher education

Standard	Features of quality scientific level
Teaching staff	understanding the needs of students
	regularity in the education process
	commitment to scientific method
	development of intellectual skills
	development of national sense
	development of analytical trends
	development outlook depth
Scientific method	cover the basic subjects
	proportionality with the ability to accommodate student
	linking with scientific reality
	basic knowledge
	learning a foreign language
Scientific reference	scientific level
	directed the scientific reference
	the availability of scientific reference
	the authenticity of scientific article
	directions provided by reference
	personal interaction awareness of the role of scientific capacity
Evaluation method	objectivity and consistency
	objectivity and comprehensiveness

	focusing on the analytical capacity
	focusing on critical thinking
Rules	availability of the necessary information
	orientation towards the labour market
	efficient and effective administrative system

The introduction of a professional system to ensure and improve the quality and excellence requires the establishment of national bodies, semi-independent quality assurance, quality improvement in the performance of institutions of higher education and improving the quality, access of higher education in Iraq, therefore it is suggested the following: The establishment of a national center for the development of university education.

• The promotion of scientific research in universities in Iraq.

• The establishment of a joint body for cooperation and coordination between all representatives of the labor market.

• The achievement of contemporary concept in education.

• Increase the percentage of acceptance.

• The establishment of the National Center for guidance.

Ranking the Iraqi Universities

The aims of Ranking Iraqi Universities are: the assessing of Iraqi universities among themselves, assessing the gap between Iraqi universities, increasing competitiveness among Iraqi universities assessing academic excellence of each university. The Iraqi universities need to watch their competitors in managing their activities, positioning themselves in higher education sector, building competences for future, and allocating resources. The ranking of universities requires universities to develop competitive strategies to gain and sustain competitive advantage in the higher education sector.

Universities in Iraq are those that published articles in the period 2003-2008 in journals covered by the institute for scientific information (ISI) in science citations index (SCI) science citations index expanded (SCI-Expanded), and social science citation index (SSCI).

Reform of university usually precedes the beginning of Governments that have economic programs based on the trend to market economics. The work forces of supply and demand are an alternative to the process of central planning. Promotion of the private sector, create jobs opportunities, bring real advanced technology and increase exports in line with the general trend prevailing, for the time being, which finds the country administration of economic activity must be minimal and therefore must reconsider the role of the public sector and size.

The reform program for included university such as: (partnership) is based on the impact, vulnerability, and joint mutual work between two or more universities related to the same goals, plans, standards and values.

Partnership has become necessary to postmodernism, and has given the universities a role to transform communities for learning. This partnership is characterized by multiple images to conduct research interfaces between different faculties and departments, implement projects for the benefit of research institutions, service productivity and others in the community, as well as providing different pieces of advice.

The partnership also represents the essential input for the conversion of universities to learning communities. This partnership contributes in improving the performance level of the university prestige, and scientific communities which is led by the universities, to improve the performance of social communities and solve great number of problems.

The reform of curricula becomes a necessity if higher education institutions in the developing countries produce graduates who are able to participate and compete in the progressing countries. In many cases, failure of reform is due to the use of the bureaucratic style, and must therefore encourage all interested parties to reform such as: student's administrations and business owners to express their views.

Economic Profiles of Higher Education

Evolving economic and social development through what is being offered of educated manpower, result a relationship between higher education and the economy. If we examine higher education, we find a combination of consumption and saving where the family is spending on higher education for future profit.

On the other hand, there is a set of economic challenges related to higher education that can be considered a package of economic and educational problems, including:

- The role of capital.
- Preparation and qualification for work.
- The emergence of new disciplines.
- Higher education funding.
- Diversity of sources of higher education.

Capital representatives of the growing role of market regulations at the university may lead to "the end of the university" and thus the loss of independence, resulting in replacing the rule of the rise of economy standards. The rise of administrators in the face of academics, the dominance of large companies and capital to work and higher education become   a captive market system.

Higher education is at the forefront of the factors responsible for the economic growth experienced by developing and developed countries for the existence of technical skills, professional and specialized disciplines, where companies depend on selecting employees carrying scientific degree, and those who would perform the responsibilities on time and pass examination in unusual circumstances and adopt these data as an indicator. The test also forms a positive impact of education on economic growth.

Concerning the increasing of the quality of the work achieved, due to the positive impact of higher education, the "Asian Development Bank" in 1989, found a strong correlation between the number of years of education in the early eighties of the twentieth century and the annual rate of change in the "GNP" (General National Production) for everyone during 1965-1985 in the economy of thirteen developing Asian countries except the Philippines and Sri Lanka.

Some believe that ignorance of higher education stands behind it. This opinion is adopted by the economists who determine a simplified method for assessing the return on investment in higher education. The fundamental problem is the measurement of the return on education during the distinctions of other.

Higher education achieves the benefits of outweighed increase in income which receives an academic degree, such as participation of higher education in leadership and management culture. For example, it was noted that the export of higher education services contribute greatly to the prosperity of the United States economy in 1999. The United States gained the largest sources of higher education services in the world.

The investment in higher education represents the budget spent on scientific research. The UNESCO institute for statistics report in 2004 indicates two things: the first is Total expenditure and second is Expenditure intensity. Both are increased in developed countries, and decreased in developing countries.

Universities in the world

The reaction of traditional university is in an attempt to restrict the spread of distance teaching and reduction of liberty. These universities of learning represent highly technical institutions, working in a team to develop their potentials through the production of knowledge; yet to follow the rules and regulations facilitators have to work by using advanced technology in the learning process. This team is supposed to be balanced between being scientific and social. It should also be open to the outside world.

The composition of this community is carried out through the organization of the university students requested in small groups with the supervision of teaching. These communities may appear by default and called upon the educational composition of the default and then via the internet. The advantage of this free society produces for its members, through exchange of experience, a constructive dialogue and reflection in force. The exchange of experience and knowledge, and therefore the virtual community, is opened to all individuals of all societies in the light of growing knowledge capacity of associates to improve their knowledge and skills and patterns of thinking.

The conversion of universities to learning communities, is required to learn the contemporary methods of learning, active learning, constructive learning, cooperative with the integration of contemporary techniques and methods of research, investigation and coordination with continuous learning and dissemination of valuable intellectual culture.

Some universities use the pending circuit television broadcast program from within the traditional university, and some colleges of America are based on the study association. The universities in capitalist countries including the U.S., Britain and France are many and varied. They are largely affected in their early opening by German and British universities, in growth and quantity. The degree that has comprises over 40% of students of secondary education is a high percentage compared with other countries. Institutions of higher education in America are classified to four main types of small colleges, technical institutes, liberal art colleges, and universities. In turn, they are divided into two types: government institutions funded by federal government, and another that receive private funds.

The British universities are estimated to be at about fifty universities. Oxford and Cambridge are the oldest established ones. They were founded in the twelfth century. The method of teaching in these universities is called the Pilot Tutorial Manner. There are other universities such as: Wales, Edinburgh, Aberdeen and Glasgow. Universities of British traditions stemming from the universities of Oxford and Cambridge have changed to some extent. The development of social changes has covered many types and patterns of higher education.

Law of education reform in Britain in 1988 became a turning point in the history of education according to the British following criteria:

- Broad authority to the Minister of Education.
- Control of the academic freedom.
- Development funding council for the financing of university of academic activities.

The French Universities are regarded as centers of university traditions, including the University of Paris and a number of universities around, supervised by the Ministry of Education. The reform law of French universities in 1968 is characterized by:

- Reforming and developing the curriculum and examination systems.
- Involving students in university administration systems.
- Organizing universities on the basis of units of education and research.

There are different types of university education in France such as the traditional state-funded university of. The three phases represent the following:

- The first phase

Joining the University for two year Degree and Diploma granted by the university.

- The second phase

Represent two separate years:

First year leads to license, the second leads to Master.

- The third phase

Postgraduate Master leads Doctorate of third link.

University Doctorate regulated the same university and Doctorate state grants for scientific research with the original level.

The universities of the former Soviet Union were of three types, including:

- Universities
- Technical Institutes
- Specialized Institutes.

These universities grant various degrees, including the first university degree. High Diploma Studies take between 4-6 years degree courses, the equivalent of science Doctoral degree that requires three years of study. The Doctor of science, which is the highest degree, was issued in a decree in 1990 to reform universities with greater autonomy and broader authority for universities.

The Arab and international experiences in the field of E-learning at the university began in Education in the year 2000. Characterized as an important educational issue in our contemporary terms by giving to science students the ability to search and investigate and find modern information. It also confirms that the scientific learning does not mean merely exploiting the potential of modern technology in the delivery and knowledge. Students can now attend the university located in places other than those they live in. Accordingly E-learning includes concepts, philosophy, objectives, certain patterns, pillars, requirements and mechanisms.

In order to shed light on distance education and its importance, it is suggested the following points:

- The reality of distance education in terms of goals and ambition.
- The importance of the challenges facing the movement of distance education.
- Effectiveness of distance education compared to the traditional model of education.

It is also to stresses that the Arab strategy for distance education in 2004 was to:

- Liberate human beings rights and maximize its contributions to progress of development.
- Meet the need for integration with other learning systems, including the education sector.

The distance education system is an educational model based on modern technology, but there are many obstacles facing the problem of application, such as: recognition, criteria, measurement, standards and quality and others. The positive results on the importance of distance learning as model of education compared to other models, suggest thinking in different forms and regulations and renewable methods of distance education to suit the point in time of the third millennium.

Adoption systems of distance education through computer networks on the concept of the overall approach include a series of general educational measures in electronic form so as to be accessible to scholars. Distance education is one of the important means of communication and technological revolution in the transfer of knowledge. It is used to develop human capabilities and enable environment for communication with the world of technology and information between individuals and among all sources of knowledge everywhere, and produces the network of the learner with constant direct contact with science. There are also regularly available information, photos, and recordings through the web, along with holding meetings and symposia.

The Ministry of Higher Education and Scientific Research with the American Academy of Science to convene   a special joint program, to develop university libraries and to qualify Iraqi digital age, to bridge the information gap experienced by universities and securing sources of information more sophisticated to either academic researchers or graduate students. This information includes teaching and working on rehabilitation of university libraries and to familiarize them with modern information technologies used in libraries of major scientific institutions of academic science which is called virtual library.

The Open University appeared in Britain in1991 within the philosophy and objectives of the core functions of education, scientific research and openness to social surroundings. It is a kind of open study, because of the inadequate equipment, rooms, and laboratories. The idea of the Open University did not receive its application in Iraq and other Arab countries, except Palestine. The Open University has resorted to innovative teaching methods, based mainly on self-education exercised by the students themselves not by studying books and educational materials that are registered by the Open University, and through follow-up radio and television programs broadcasted regularly in terms of appropriate information.

The use of the Internet has become a reality. It has imposed itself as a modern method used in university education, but there are many things to identify clearly the benefit from its use and encourage its spread in future, including:-

- Descriptions of output achieved by the kind of education.

- Clarification the methods of coordination between various education systems through the network.
- Development of assessment methods appropriate for students.
- Update instructions in line with the software used in line with the software used for this type of education.
- Cooperation colleges in the corresponding decisions similar to this type of education.
- License modernization so as to protect the rights of electronic publications.
- Determining academic standards for admission.

The activated education addressed to stimulate skills, to attach a variety of potential users, to remain with them throughout their lives, to take responsibility for themselves and their learning. It is evident that such constructive learning environments through successful players should be well prepared containing different process, and the link between qualified teachers with a renewed mind movable teacher has features of mobility.

Contemporary Problems of Universities

The negative contemporary universities problems are inherited, while positive changes create a lot of intellectual, cultural, social trends as well as the contemporary university job to prepare manpower and frontier science and scientific research, cultural and intellectual heritage.

Academic disciplines have evolved with the development of sciences and various new disciplines such as engineering, agriculture, science and total treatments that began in the nineteenth century, whereas the twentieth century indicates other disciplines such as business management, journalism, information and library science, economics, politics and world affairs were added. Each country has special methods to determine its own disciplinary university and identification numbers, graduate students and the quality.

The world witnessed in the twentieth century   a breakthrough in all fields and scientific trends, so there are no boundaries between different disciplines. For example, medical science requires engineering science and recent tests of modern science depends on the physical, chemical extraction and analysis and also relies on mathematics to lay the groundwork mathematics   .

The progress and development in pure science, for example, develop new subjects and disciplines and specialties of science. New interfaces were not known during the first half of the last century. The results of these major changes in curriculum and build up research transformed these developments to the university curricula. Seminars and researches are now carried out in different ways including:

- Bachelor degree is based on the study and thesis.
- Some universities in Britain and Germany developed curricula at the level of initial studies that include research and study.
- Division of the present fields such as industrial chemistry, chemistry of life with medical side and other disciplines in the branches of pure science.
- Development of competencies.

These terms of reference have been developed in American universities in physics, chemistry and mathematics, to prepare graduate in some sectors such as engineering, chemistry physics and chemistry, agricultural engineering, and medical studies.

- Addition of assistance topics.

Some topics have been added as assistance of many of the terms of reference of pure sciences, including education, literature and library services and use of modern machinery and computers.

- Other disciplines (Sandwich)

British universities were carrying out by expanding the initial years of university study for use in increasing opportunities for the systematic teaching and applied for and rehabilitation work in various production sector.

Chapter Five

Development of science of biotechnology sector /the Iraqi perspective

Preface

Iraq occupies a very diversified geographic and climate zone and assumes a great deal of importance. However, its wealth is under U.N. resolutions. Its health and agricultural problems are diverse.

For almost twenty years (1970-1990), its Biotechnological activities have sought to bear on various problems on Agriculture and Medicine. In recent years (1990s), these activities have turned in a very limited way, to use some of new developments in Basic Biotechnology such as Molecular Biology and Genetic Engineering.

Various Centers and Departments are involved in the National Biotechnology research systems in Iraq. The Biological and Agricultural research Centers also integrate such activities into their research program.

Presently, these centers established different types of programs (long term, short term) in Agriculture and Medicine, for generation, development and transfer of Biotechnology. Major components of these programs; include development of technology transfer facilities and establishment of international collaboration in the field of Biotechnology.

Medical Biotechnology in Iraq, as a whole, demands strong collaboration in order to reduce the present gap between Iraq and the developed countries. Whereas, Agricultural Biotechnology seeks to increase food crop resistance to insects and diseases, utilize a wide varieties of techniques, including Biological control and various forms of Biotechnology, such as tissue cultures, molecular genetics and genetic engineering.

This part of the analyses of the major activities of Biotechnology in health, Agriculture and Basic Science in Iraq offers general background information on the development of Biotechnology. Then examines the current of Biotechnology research and potential, and suggests an outline of the strategies for the future of Biotechnology in Iraq.

Iraq, which has already entered the biotechnology era, whether on purpose or by chance, will in any case inevitably suffer the consequences of the widespread use of such techniques in industrialized countries. Therefore, the question is no longer whether such a move is desirable. but more how can best be put into use.

Currently people of Iraq are suffering from poor quality of life because of U.N. sanctions, which resulted in:

- Food shortage and malnutrition.
- Unsafe drinking water.
- Improper sanitation system.
- Poor health care.

This is why Iraq needs to have at least a basic level of skills to make it possible to define and implement policies in Biotechnology and the other technologies. Biotechnology alone will not feed Iraqis or give them better health, but it would be irresponsible not to see it as one of the available tools. A certain number of social, cultural or institutional conditions also have to be satisfied before technical success can be converted into economic and social progress.

This part analyses the major activities of biotechnology in Iraq; then examines the prospects of strengthening biotechnology in conjunction with conventional technologies; and finally it outlines the strategies for the promotion of biotechnology in Iraq.

Historical Aspects of Biotechnology in Iraq

Biotechnological activities were known to ancient Iraqis in Samaria and Babylon. For instance, fermentation processes and perfume extraction were very well-known to them.

It has been reported that the first experiment in scientific research was performed in the palace of one of the kings of Babylon. The experiment was carried out by two palace ladies who prepared perfumes by distillation from plants (steam distillation).

Cheese making is believed to have started in Mesopotamia at approximately 8000 years ago.

In the early 1970s scientific infrastructure construction began and several research activities were planned and set up.

Most of these biotechnological activities were limited to traditional methods to serve their needs; i.e. industrial fermentation, soil microbiology and bioconversion of waste products.

The government showed its interest in providing support to biotechnology by offering to host the first Arab Conference on Genetic Engineering in 1984. As   a consequence of the meeting, the council of Scientific Research established "Genetic Engineering Center", which became responsible for all research and development in this field; then in the agreement that Iraq established an affiliation of the International Center of Genetic Engineering (ICGEB) with the Scientific Research Council, as the principle Liaison Institute.

Since then during early 1990s research in agricultural biotechnology dealing with production of Biocides and Biofertilizer and Tissue Culture Technique were carried out by (IPA) Center.

The Ministry of Higher Education in the middle of 1990s, established a new centers for Genetic Engineering, involved in research on biotechnology. The main research objective on biotechnology of these centers involved the application of basic genetic engineering particularly in the following fields:

- **Molecular biology**: for enzyme production, industrial and clinical diagnosis.

- **Cell biology**: for animal cytogenesis (chromosome and gene mapping).
- **Microbial genetics**: for the production of useful microbial compounds through genetically improved strains.

Iraq through a new council of biotechnology at the University of Al-Nahrain (previously Saddam University) is monitoring international development in biotechnology in order to apply some of these techniques. As a result, this council has developed plans for biotechnology and genetic engineering in the sectors of health, agriculture and basic biotechnology. The major components of this plan are:

- Generation, transfer and development of biotechnology.
- Development of trained personnel.
- Strengthening of research and technology transfer facilities.
- Establishment of international and regional collaboration in the field of biotechnology.

Approaches of biotechnology in Iraq

Most biotechnological activities, which are applied in Iraq, are limited to traditional methods and serve their needs through:
- Fermentation
- Antibiotic industry
- Single cell protein
- Plant biotechnology
  - Tissue culture
  - Soil fertility, through biological activities
  - Increasing food production through plant cell-culture
  - Bioconversion of waste for food and feed ingredients

Fermentation

Iraq in early 1970s established traditional fermentation industries, bakers yeast production, ethanol, acetic acid, acetone, butanol and citric acid production. In 1970, a factory for making bakers yeast from sugar-beet molasses was established with plans for the production of compressed yeast.

This industry was faced by many problems, especially in dried yeast production:

• There are abundant sources of raw materials for fermentation in Iraq. Large quantities of hydrocarbons, and carbohydrate by-products (molasses) and lignocelluloses waste are found.

• Fermentation of food crops like dates, which enjoy comparatively large market sizes, does not receive sufficient attention. Local research on bioreactors is therefore needed to support the development of new processes and improve the performance of food industry in Iraq. Bioreactors also play a key role in the production of enzymes used in the beverage, detergent and leather industries.

Antibiotics industry

The antibiotics industry in Iraq started in 1970 for the production of penicillin and tetracycline. On the other hand, tetracycline production continued until 1980. This industry was discontinued for economical and technical reasons as a result of sanctions. The involvements of research and development programs have positive effect for restarting bio-industry and production of different types of antibiotics.

As the pharmaceutical industry in Iraq is directed to satisfy the local market by satisfying the market needs, it is therefore necessary to develop appropriate biotechnologies against various diseases that are endemic.

Single-Cell protein

A research and development program for single-cell protein (SCP) production at pilot plant level started in Iraq in 1982. The production of SCP from methanol using local and imported strains of Candida Utilis was investigated. The research plan included an economic feasibility study and the assessment of technological and nutritional aspects of SCP under local conditions. Then, another pilot plant was established to utilize date syrup for the production of bakers yeast and SCP.

The most important achievements of this program were the establishment of pilot research facilities, training of personnel, the nutritional assessment of available commercial SCP products and the isolation of several methanol-utilizing bacterial cultures.

Plant biotechnology

Plant biotechnology applications in Iraq include soil fertility (nitrogen fixation) using yeast strains in a mixed culture and cell conversion. Scientists have carried out research on nitrogen fixation by grain legumes. The main objective of this research was to increase the yields of grain legumes while decreasing the input of inorganic nitrogenous fertilizers

Tissue culture

• In view of the importance of the date-palms in Iraq, a tissue culture laboratory was established in 1979 at the Agriculture and Water Resources Research Center in Baghdad. Another laboratory was initiated at the Genetic Research and Biotechnology Scientific Research Center for the improvement of plant production. Tissue culture laboratories were also established at the Universities of Basra and Musol and also at the Ministry of Agriculture. The latter worked in collaboration with FAQ through regional center for date-palms in the near east and North Africa. The major objective was the commercial propagation of plants by in vitro techniques.

• Then in 1982, the vegetative micro propagation through tissue culture was carried out as a promising technique. However, future research is needed to early following and lack of uniformity of the closed plant.

• Date palm propagation by tissue culture was also implemented at research institutes and universities in Baghdad and Basra. Other species such as lettuce and potatoes were propagated by tissue culture at the department of Biology of Mosul University.

Other activities

• Research at the Faculty of Agriculture and Biology in the Nuclear Research Center was concentrated on the study of a sexual embryogenesis techniques and the induction of mutations by mutagenic agents.

• Agricultural and forest residues in Iraq are considered renewable resources that can be utilized by bio-technological means for the production of food, fertilizers and fuels.

The situation of Biotechnology in Iraq, since the imposition of sanctions

As a result of sanctions, many fields in public health system are affected sector by sector including critical areas such as: food security and nutrition, water resources, women's health, children health, national health emergencies, hospital care, humanitarians' donations and international cooperation. Scientific fields are also affected such as Oncology, Cardiology, Nephrology, Endocrinology, Ophthalmology, Diagnostic Testing and Protection of Blood Supply and Scientific Information and medical education, pharmaceutical and biotechnology inputs.

In agriculture, the U.N. sanctions ban the importing of fertilizers. Shortages in production of crops led to the deterioration in the Iraq's populations and nutritional intake (daily caloric intake).

• Iraq is facing the following three main threats: food supply, health improvements and environmental protection. Millions of people in Iraq under poor and risky growing conditions are suffering from poverty and poor health. They go to bed hungry due to U.N. resolution and security problems.

• Food security in Iraq is unique; it is not related directly to biotechnology problems. Genetically altered seeds are not necessarily needed to feed Iraq. This view rests on two critical assumptions, which we question:

The First is that poverty is not due to a gap between food production and growth of population.   **The Second** is that biotechnology is not the only or best way to increase agriculture production.

- Iraq is not able to undertake effective agricultural biotechnology research for its own urgent needs without the scientific support of developed countries. Iraq needs more investment in developing appropriate agricultural biotechnology.

- Iraq which is subtropical country is basically an agricultural country. Approximately 80% of the total area is devoted to cereal production mainly wheat, barley, rice and corn. The remaining 20% includes a wide variety of crops, date-palm, citrus, tobacco, cotton and others. In general the average production level of all the important crops in Iraq is very low compared to those in developed countries. In the last 10 years, plans have been organized by the Iraqi government with cooperation of agricultural scientists to improve both the quality and quantity of crops per capita, using proper machinery equipments, fertilizers and suitable control measures.

Applications of biotechnology in Iraq in the experimental stage

Animal production

- The use of biotechnology in animal production in Iraq has occurred, in the field of reproduction, animal health, feeding, nutrition, growth and production.

- In the field of reproduction, new biotechnologies such as embryo transfers, in vitro fertilization, cloning, and sex determination of embryos have been studied experimentally for different types of livestock at the faculty of Agriculture and Biology in the Nuclear Research Center.

- Animal health can be improved with new biotechnology methods at experimental stage of diagnosis, prevention and control of animal diseases.

- Biotechnology in animal nutrition concentrates on improvement of feed; enzymatic treatment, the decreasing of the anti-nutritional factors in certain plants, such as legumes, which are used as feed.

Plant production

At present, more traditional aspects of biotechnology such as the followings are used:

• Tissue culture

The application of tissue culture does not require very expensive equipment. This technology was applied in Iraq to improve local varieties of food, crops for example using traditional methods for propagating potatoes for example.

• Pest and weeds

In Iraq, most of the land is affected by many weeds causing big losses in the agricultural crops. Since 1970s several herbicides were used to control the weeds of corn, cotton and vegetable fields.

Bio-control programs used for controling pests

The plant production research center is a State Board for agricultural research. College of Agriculture, Baghdad and Mosul Universities, Agriculture and Biology Research Centers and Iraq Atomic Organization are the main research centers in Iraq, which carried out different research studies on the biology, taxonomy and control of the pests. These research centers successfully adopted control measures on the most important pests attacking crops, vegetables and by applying chemical and agricultural methods. Promising results were obtained from many plant extracts against pests, by inhibition pest life cycle.

In Iraq, most of the agricultural land subjected to grow many of weeds causing big losses to the agricultural crops, as many research workers confirmed the positive results of the weed control measures to increase the yield of the crops.

Researchers started to evaluate herbicides since 1965, and mid of seventies, herbicides were applied to control the weeds in wheat, rice, corn, cotton, potatoes and tomatoes. Increased support is needed to expand research designed to develop new herbicides that are not likely to pollute ground water and that will provide reliable control.

Several centers are involved in bio-control programs.

commercially, research centers introduced two bio-control mutant fungi. Both fungi successfully were applied to control plant parasitic nematodes and soil born fungi on vegetables and citrus. Also the center used another fungus against date-palm stern borer insects.

State Board for agricultural research successfully adopted several control measures on the most important pests attacking field crops, vegetable and fruit trees by applying chemical, biological and agricultural methods. Many insect growth regulators, bio-control agents, fungi and plant extracts were experimentally applied on small and large scale fields.

Most of the research studies of the graduate students in the department of plant protection concentrated on the biological control, and plant extracts.

At the present time, the U.N. sanctions which have been imposed on Iraq since 1990, destroyed most of biological control programs and Iraq is facing lack of well trained personnel and shortage in facilities. As a result the first generation biotechnologies used in Iraq such as insect resistance, herbicides resistance are not easy to address any more.

Health biotechnology

- Several centers are interested in carrying on research on health biotechnology. The following may be mentioned.
- Previously Saddam Center for Cancer and Medical genetics Research (SCCMGR).
- Institute of Biotechnology and Genetic Engineering for Graduate Students, University of Baghdad.

- Department of Genetic Engineering, College of Science, University of Baghdad.
- Genetic Engineering Departments in a number of Universities.
• SCCMGR is engaged in several lines of biotechnological activities in health:
- Cloning of tetanus toxic gene into tumor cells.
- Preparation of tumor cell lines in vitro for gene therapy technique
- Preparation of restriction enzymes vectors for gene therapy.
- Studies on disorders of mitochondrial DNA in muscular dystrophy.
• Several biochemists have participated in various research projects that deal with diagnosis, and monitoring of several types of tumors.

Future biotechnology

The need for the application of biotechnology to face the basic needs regarding food and health in Iraq is real. There are different approaches such as the development of plant biotechnology, biotechnology applied to livestock production and biotechnology applied to food processing.

So, the suggested components of biotechnology plan include:
• Micro propagation: through e.g. tissue culture for multiplication.
• Genomics: the molecular characterization of all species.
• Bioinformatics: the assembly of data from genomics analysis into accessible forms.
• Diagnostics: the molecular characterization and identification of pathogens.
• Molecular breeding: the identification and evaluation of desirable traits in breeding programs with the use of marker assisted selection
• Transformation: the introduction of single genes conferring potentially useful traits into crops, livestock, fish and tree species.
• Vaccine technology: use of modern immunology to develop recombination DNA vaccine for control of lethal diseases.

Plans for future biotechnology research should be formulated through several priorities:

- Food security.

- Increase and improvement of agricultural production. Breeding for higher- yielding plant varieties and improve nutritive development values. Pest and pathogen-resistant genotypes and conservation of plant genetic diversity.

- Production of pharmaceuticals for the extraction of biologically active plant substances.

- Immunology: Production of vaccines and monoclonal antibodies.

- Use and recycle of agricultural products for the production of ethanol, acetone, butanol and methanol.

Food security

The applications in agricultural biotechnology in Iraq have the promise of bringing about the much-needed requirements in agricultural production, such as carrying resistance / tolerance to a biotic stresses (drought and salinity) and to provide options for better rotation to conserve natural resources.

Iraq is neither in the process of testing genetically enhanced products in number of crops, nor in the process of testing commercial products in the market to-date. Many technological advances are not visible in the farmers fields in Iraq but in the future are expected to provide ways to improve crops in a precise and fast manner. Use of functional genomics to address complex traits, marker assisted breeding to ensure presence of key genes, improving nutritional quality and managing natural resources better by use of efficient monitoring tools, Iraq must be an active participant in this area so that specific needs of food security are achieved.

Agricultural production

Despite the importance of health and industry sectors, the suggested priorities in the plan should have great emphasis on agro biotechnologies for two reasons. **First**: research on plants for crop improvement directly relates to specific ecological conditions predominant, whereas biotechnology applications for industry and human health are more difficult in Iraq. **Second**: preliminary data indicate that most biotechnology research activities in Iraq relate to agriculture.

The following classes of agricultural biotechnology are suggested to be used in Iraq in the future:

• Gene transfer technologies, which provide transgenic plant, resistant to many pests' pathogens, herbicide resistant to stress such as temperature, drought and salinities.

• Non transgenic biotechnological approaches for improving the efficiencies and effectiveness of conventional plant breeding methods.

• Technologies for better monitoring of natural resources and environment.

Additional suggestions for the implementation of the plant are the following:

Plant biotechnology

- The establishment of biological treatment plants for sanitary wastewater and utilization of the treated wastewater for landscaping and agriculture.

- The establishment of the commercial production of inoculants such as Rhizobium and developed efficient methods for recycling agricultural waste such as beet molasses and maize and rice straw. Research should be carried out on the use of bio-fertilizers to increase rice yields in Iraq.

- The use of biotechnological techniques for the development and improvement of bio-insecticide for control of plant pests.

- Increasing protein content of rice by the application of biotechnological techniques. Rice is one of top five cereals grown in Iraq with its unique capabilities desirability to grow in stress conditions it is a crop of choice for Iraqi people. Rice has received comparatively less attention for research in general.

- The creation date-palm clones resistant to disease, and the application of tissue culture to improve date-palm varieties.

- Future research is needed to overcome the difficulties related to early flowering and lack of uniformity of cloned plants.

- Production of secondary metabolites, by tissue culture, the selection of plant cell lines for stress tolerance to salinity and drought and also the production of virus- free potatoes planting material; and the micro propagation of plant.

Animal biotechnology

Biotechnological application in livestock and fish production, and the adoption of embryo culture to improve local animal breeds through embryo transfer technology, are samples on pre-implantation and embryo freezing.

Microbial biotechnology

- Microbial biotechnology for ethanol production from sugar by-products and methanol production from Agro- industrial wastes.

- Microbial genetics: elimination or degradation of pollutants transformation of cellulolytic nitrogen fixers; construction of Saccharmyces cervical strains capable of cellulose, cellobiose or lactose consumption.

- Proper technology to convert biomass into bio-fuel and biogas to convert agricultural biomass and animal droppings into bio-fuel and manure; of the various wastes used for biomass production. Rice straw is one of the possible applications.

- Bioconversion of lignocelluloses wastes to protein- enriched fermented materials followed by the production of microbial biomass from the hydrolyzed cellulose product.

- The use of bacterial treatment for the removal of oil and chromium; and also in the nitrification-denitrification process to remove ammonia .

Health biotechnology

234

Medical Biotechnology in Iraq, as a whole, will demand a strong collaboration in order to reduce present gap between developed countries and Iraq to achieve this aim:

• A center of bone marrow transplant should be established in one of the hospitals. This requires the availability of experts, equipment and materials.

• Iraq is facing cancer and genetic diseases hence, research projects in gene therapy should be initiated.

Pharmaceutical industry

The pharmaceutical industry in Iraq should be expanded and developed so as to meet at least the local requirements. Biotechnological techniques should be introduced.

Environment

- Applications of natural occurring organisms (e.g. yeast, fungi and plants) should be used to convert hazardous substances in soil.

- Using microorganisms' pollutants from sewage systems to clean up industrial sites.

- The use of biotechnology to avoid pollution is of increasing importance such as the use of bioreactors to treat hazardous products.

Bioinformatics

This new discipline (biology and computing), which will be the core of biology in the 21st century, should be used for measuring and monitoring thousands of genes at one time. This computer-aided bioinformatics will stimulate future developments in the pharmaceutical industries.

cooperation with International Agencies

• Cooperation with Islamic and International agencies and countries are required.

- Well trained scientists from Arab and Islamic countries, directly involved in the training and transfer of various biotechnologies are also required.

- Post-graduate short training courses sponsored by international organizations such as the United Nations Educational, Scientific and Cultural Organization (UNESCO) should be organized in the various fields of biotechnology.

- Training of medical personnel in bone marrow transplantation, to help in gene therapy especially for cases of leukemia and lymphoma.

- Broadening the biotechnology base is a must for characterization, collection and conservation of germplasm that is already in the gene bank collection around the world and providing information for collection of gene pools that are not currently available in gene bank.

- Strengthening capabilities, developing projects, visits, and training programs of mutual interest to all participating countries in the following areas of biological control:

  - Exchange of biotic agents on a case to case basis.
  - Mass production of host insects and natural enemies.
  - Biological suppression of crop pests by developing joint projects.
  - Computerization of information and networking research organization in different countries.
  - Training in different aspects of biological control.

Ethical issues related to biotechnology

The Islamic world needs to have sharp opinions on various current issues related to biotechnology and genetic engineering such as genomics, human cloning and genetically modified organisms.

The National Programme for the Biotechnology / proposed research projects

Bio-technologies represent the modern mosaic of knowledge, rooted in the world of genetics, biology, evolutionary biology and molecular knowledge of chemistry. Bio-technologies are unique among other technologies for being a tool for dealing with life itself. They include technical applications that use the vital systems of organisms or their components or products to modify products or vital processes for specific purposes, and therefore include many operations and have wide application in agriculture, industry and other sectors.

Despite the benefits of bio-technologies in the fields of medicine, agriculture, industry, however other biological fields increasingly intervene with many other subjects and indicate that the diversity of thought began to appear. Therefore, a significant problem surfaced when preparing the National Programme for bio-technologies. Then, there was a significant overlap between the main axes of the programme and its affiliates. It developed themes and sub- themes according to different areas of bio-technologies that can be investigated by the Iraq meaningful progress particularly in the areas of food and health and strengthening pharmaceutical industries.

Major components of the programme

- The biotechnology and food security
- The production of bio-pesticides.
- The production of bio-fertilizers.
- Tissue culture.
- The production of potato tubers and seedlings free of viral disease
- Production of broad Date Palm
- The production of Fruit assets.
- Production of secondary materials
- Education programs and improvement.

- Technical embryo implantation in cattle.
- Manufacture of food products with curative nature.
- The production of liquid sugar from Iraqi dates.
- A study of pesticide residues in food.
- Improving the nutritional values of feed.
- Single-cell protein production.
- Improving the conditions of storage of foodstuffs.

Health and medical technologies

- Gene therapy and diseases
- Markers of neoplasm
- Genetical markers
- The production of vaccines
- The production of antibiotics
- The production of hormones and enzymes
- Vaccine production
- The production of antibodies to microorganism and toxins from snake bites and insect bites
- The production of antibodies (monoclonal antibodies)
- Development or establishment of cancer lines in the laboratory
- Transfer of bone marrow and cultivation of bone marrow
- Using molecular indicators in human lymphatic cells
- Diseases of hereditary Cancer
- Diagnostic kits
- Forensic-genetic fingerprint

Biotechnology and safety pharmaceuticals

- Extraction of drugs from living organisms (plants, animals and microorganisms) for use against cancer and other diseases.
- Extracting of active substances.
- The production of pharmaceutical materials and medicines through the revival of genetically modified microorganisms.
- Preparation of medicines from medicinal herbs.

Plant Bio-technology

- Plant tissue culture, both at the level of cell or tissue or at the level of protoplast.
- Improving the quality attributes of different crops.
- Production of plants of potential environmental conditions.
- Production of plants resistant to pests and agricultural bush.
- Increased production of medical and pharmaceutical plants and medicinal herbs.
- The genetic diversity of plant.
- The production of fertilizers and bio- pesticides.
- Development of insecticides to control plant pests.

193

- Reproduction development of resistants to diseases and the application of tissue culture.
- The production of secondary materials using tissue culture.
- Production of plants that have the capacity to fix Nitrogen.
- The production of new crops resistant to pesticides and pests salinity.

Animals and microorganisms technologies

- The production of genetically modified animals characterized by the attributes of high productivity.
- Animal tissue culture.
- Improvement animal resistance to environmental conditions.
- The production of veterinary vaccines.
- Artificial insemination.
- The diagnosis of hereditary diseases and germ using the PCR (Polymerase Chain Reaction).
- The cultivation of embryos.
- Applications of biotechnology in improving livestock.
- DNA technique.
- Vaccination of livestock against diseases.
- Transmission of embryos in cattle and fish.

Microbial Bio- technologies

- The production of hormones
- Fermentation
- Producing bacterial strains of anti- cancer
- The production of bacterial strains of high productivity of enzymes
- The production of bacterial strains of high productivity of antibiotics
- The production of microorganisms strains which have desirable qualities for use in bio- pesticides and fertilizers
- Producing microorganisms strains with high productivity of industrial materials
- Ethanol production by microbial Bio-technologies from sugar and methanol production of industrial and agricultural waste
- The removal and crushing and transformation of nitrogenous cellulosic fixtures
- Bio-conversion to cellulose waste materials rich in nitrogen

Environmental Bio- technologies

- The Use of living organisms in purification of heavy metals of the environment.
- Preparation biocides to combat agricultural pests.
- Preparation of bio-fertilizer to improve agricultural product.
- Finding microorganisms to revive the disintegration of some hydrocarbons and turned into a simple compound.
- Bio-technologies in the ecological balance.
- Bio-treatment of sewage.
- The use of microbial treatment to remove oil, chromium and ammonia.
- Using natural organisms to convert hazardous substances in the oil.
- Clearing industrial sites and avoid pollution.
- Oil Pollution by breaking chemical compounds.
- Isolation of bacterial strains that have the ability to remove sulfur.
- Using Bio-Markers and Bio-controls for detecting pollution levels.

Bio-technologies of Water

- Aquaculture techniques
- Techniques of bio- actors of water
- The use of bio- sensitivity
- The use of drugs and vaccines to preserve the health of fish.
- The cultivation of fish.

Biotechnology and basic sciences

- Genetic engineering techniques
- Genetic fingerprint (D.N.A. Finger Printing)
- Production of restriction Enzymes
- Production of standard D.N.A.
- Diagnosing the production of certain genes
- Amplification of genetic materials

Genetically modified organisms

- Genetic manipulation of farm animals to produce therapeutic human proteins
- Methods of genetic manipulation of farm animals
- Genetically altering poultry for the production of therapeutic proteins
- The use of animals in the production of genetically therapeutic proteins 195
- Adverse effects on animals that may arise due to genetic manipulation
- Genetically engineered animal, models for human diseases
- Genetically modified plants
- The production of much meat with low fat
- Improving the quality of protein from milk cows
- Gene transfer techniques to provide resistance to many of the insects and plants resistance to high temperatures and drought salinity

Bio-technologies in industry

- The production of antibiotics
- The production of enzymes and drugs
- Production of energy materials such as ethanol and methanol and acetone
- The design and analysis of Bio- reactors

Transfer Bio-technologies

- Technique of PCR (Polymerase Chain Reaction)
- Technique of PCR- STP (PCR with Short Tandem Repeats)
- Other techniques

Cloning

- Cloning techniques

Genome (genetic content)

- Studying the genetic content of the bacterial isolates
- Isolating and purifying D. N. A.
- Determining the content of the Plasmid isolates
- The safety of animal products and genetically modified organisms
- Building gene maps of plants of economic importance

Studies related to bio-technologies

- The systems and regulations for genetically modified animals
- The importation of transgenic animals
- Economic feasibility of genetic modification in animals farm to produce therapeutic human proteins
- Controls systems in the use of genetically modified organisms
- Ethics and social values and bio- technologies
- Bio- informatics

- The Bio-safety and bio-technologies
- Laws regulating the use of genetically modified crops
- Medical, religious and security considerations of bio- technologies

Development of scientific incubators in Iraq

To introduce technological incubator in Iraq it is proposed to bring these ideas systematically in line with the global approach by doing the following general functions:

• Absorbing the output and achievement within the country at the level of master and doctorate degrees and transnational consulting offices of these colleges.

• Absorbing the achievements of research and development centers in different quarters in the state (achievements of pharmaceutical, veterinary, programming and packaging materials)

• Employing the achievements of some companies from the public sector-private or mixed-dealing with specific technologies of the functions of these products, including incubators, and feeding of the proposed technologies as well as industrial products.

These technological incubators cover:

- Biotechnology
- Technology of drugs, medicine, veterinary medicines, herbal medicines, medicinal agricultural materials
- Technology of new materials, packaging, canning
- Information technology
- Technology of food industries

The importance of technological incubators in Iraq

To support the private sector, these is a clear desire to encourage this sector and meet the urgent need of the Iraqi people, especially in the sector of pharmaceutical, medical, veterinary, and food industries, packaging, information technologies, means of education and industrial sector.

Elsewhere, there are efforts to link interaction of institution of higher education, scientific research and technical institutes with productions, services as well as benefit from the ideas of creativity and innovation among individuals and institutions of Iraq. The embodiment of these trends has issued new laws and established special institutions that have national initiatives to achieve the required states of the initiative "cooperation mechanism" between the colleges and universities with various ministries and increase the number of graduate students, particularly at the stage of doctorate.

This shows that the economic climate and scientific activity and legislation in Iraq encourage the development of technological incubators at the present time (to be on   a trial basis), and the objectives for that are the following:

- The first objective is to gain experience in how to achieve Iraqi "product development" or how to move from search result to the investment, or how to transform idea and innovations or developments and renovations to the scientific and technical institutions to factories or services. This experience and expertise that produces them, if successful, must be repeated in dozens of places and scientific and technical areas.
- The success of the idea of incubators leads to diversify the Iraqi economy greatly, especially when it beats the experience of being beyond generating tens of incubators in Iraq.   The success of incubators leads to achieve substantial added value in production processes and services, not only in limited profits, leaving the added value of large foreign companies, the concessionaire or have intellectual property rights.
- Generating jobs and real productive for Iraqis, especially their graduates.
- Lifting returns of laboratories, equipment of universities, research centers and industrial development during the period of its life produced by

investing in the work for production and service sectors and thus improve the conditions of their investment.

- The success of the experiment, solving these problems and constraints encountered will transform the experience of adoption by large companies in Iraq. Technology incubators associated with these companies will lead to significant results in the generation of existing industries (down stream industries) as well as in creating nutritious industries of these companies by generating (upstream industries), and feeding industries.

- The technological incubators will benefit in the marketing of output of scientific and technical universities, research centers and industrial development and linked to the national economy more deeply.

There are many justifications for the introduction of such regulation of such regulation, because the economic situation and its structure include more coordination, such as the following:

• Supporting the industrial sector in charge of the proposed technology incubator, including (medicines, information, food industries).

• Supporting the private sector which deals with these technologies to invest in the industry.

• Contributing to the provision of products for purchase of Iraqi citizens within its possibilities.

• Providing services not normally available.

Starting with incubators of this kind goes back to the physical possibilities available as well as the provision of appropriate venues as well as the fact that the process is not easy for the existence of many obstacles, yet this is a new experiment in which many elements have been evolved.

The private companies of special paper, cardboard, plastic, aluminum sheets, the National Center for Mobilization and packaging at the Ministry of Industry, conservation and packaging supportive of the pharmaceutical and food industries (dates, vegetables, and fruits) can play many functions.

Moreover, technological incubator in Iraq could play a special assignment, including:

- Linking universities and research centers in the industry.
- Marketing output of scientific and technical communities.
- Transfer of technology from home and abroad to invest in production and service sectors.

Developing job opportunities for graduates.

- Serious desire to support the industrial sector and agriculture and the private sector.
- Generating companies in the feeder industries and expanding in the market and- added value and the generation of new industries.
- Increase the added value in production sector.

Specifications of technological incubator

The establishment of technological incubator in Iraq should be connected to the Ministry of Higher Education, to meet the urgent need for the existence of an institution that play a role of assistance and contribution in providing advisory services to medium and small- sized enterprises and individuals. Incubator package of facilities and support and consultative mechanisms can be provided during the period or periods of time until the relevant qualification is set to start production and actual work.

These incubators specialize in general successes of the features and services including:

- Medicines and Medical herbs.
- Medical and veterinary supplies.
- Informatics area, computer software and the Internet.
- Packaging, preservation, packaging (materials technology)

The private sector will participate in the sectors of pharmaceutical, food industries and information technology sector and other feeding industries and thus contribute and support the private sector in an orderly and controlled manner both in scientific and organizational transformation.

Number of international bodies and institutions, including the Economic and Social Commission for Western Asia (ESCWA), as well as UNDP, other international institution, regional, will also participate in this effort.

The mechanism of this incubator, include the followings:

• Development of training programs and consultations followed by the selection of leading scientific wish to begin work in establishing a yield of profits.

• Coordination of incubators and then choosing them from among enterprises.

• The incubator during the incubation provides financial services, legal advice and support and develops plans around the dilemma of funding and the necessary investments.

Objectives

The most import objectives of the incubator according to the proposed technologies (medicines, food industry, IT … etc.)

• Helping graduates of universities and higher institutes to establish their institutions and their own business.

• Helping researchers to use the results of research carried out in technologies mentioned from the stage of laboratory work to the stage of practical application view production.

• Contributing to the resettlement of imported technologies and to assist in the transfer of technologies from developed countries.

• The incubator in the later stages to provide advisory services to the beneficiary institutions at work sites.

• The incubator provides advice in areas such as financial budgets discretion, and funding requirements needed to start production and organization of loans and payment methods.

• The incubator in the event of the availability of funds, provide soft loans for small enterprises.

• Incubators work mainly on developing a special relationship with local institutions, global development-related administrative and transfer technologies to local universities and research development.

• The incubator is usually implemented on intensive training courses for institutions incubated on some issues related to the success of the project.

• The incubator provides guidance on the new procedures and applicable laws entrepreneurs.

Management

The incubator is linked administratively by coordinating formulas with the National Commission for transfer of technologies and by the Iraqi universities and units that are consistent with the proposed technologies, including:

• Research Unit of drug/ college of Pharmacy / Baghdad University.

• Unit of common diseases/ college of Veterinary Medicine/ Baghdad University.

• Unit of medicinal plants/ college of Pharmacy/ Mosul University

• Unit of hemoglobin morbidity/ college of Medicine / Mustansiriya University.

• National Center for diabetes treatment and Research/ Medicine / Mustansiriya University.

• Center of blood diseases/ college of Medicine/ Mustansiriya University.

• Euphrates Unit for Research on Cancer/ college of Medicine/ Kufa University.

Those emanating from the other Ministries, including:

• Centre for the pharmaceutical industry.

• Center for Research and production of the diagnostic kits.

• Research Center for the veterinary medicines.

• Research Center pharmaceutical industry.

• National Center for Mobilization and packaging.

Therefore it requires the use of materials and instructions of existing laws to put the foundation or rules and procedure of the incubator from the laws we have mentioned above.

Moreover, passing laws may require the recall in the following basic features:

- Financial requirement of legal status and rights of individual property.
- Administrative requirements.
- Right of workers in the public sector when the loan was nurtured and how to leave his work in the sector.
- Issues of intellectual property rights for products.
- Legal features of the development of fund for incubator companies as they enter and when graduated and expanding production lines   . Planning methods of benefit from the potential laboratory in the public sector and private sector.

The steps to implement the project

- Creation of a functioning business for the pioneers, examine the studying the legal aspects.
- Developing a plan of financing.
- Developing rule of procedure of the incubator.
- Agreement with the concerned parties.
- Organization of national symposium.
- Selection and training of nursing staff.
- Preparation of construction.
- Media and marketing
- Networking
- Follow-up performance

There is a number of proposed parties to finance this incubator project, including:

- Fund Development Planning Commission.
- United Nations Development program.
- Proceeds obtained from the 5% of the profits of companies that are allocated for research and scientific development.
- Cooperation mechanism.
- Incentives for creators.
- Consulting offices.
- Personal contracts for university professors.

Rules for admission to the incubator

The economic environment in Iraq determines some important rules to accept the products that will be adopted in this incubator:

• Product that has a relationship with proposed technologies that is currently marketing in the country (pharmaceutical and food industries … etc.).

• Product that has the raw materials within the country.

• Product or service with high added value.

• Product that was developed inside the country (diagnostics kits, drug … etc.).

• Product leading to the transfer of new technology (biotechnology, genetically modified organisms, genetic fingerprint)

New trends and phenomena
Of and scientific Incubators

Knowledge economy represents a new type of economy, different from the old economy, which was based on land, labour and capital as factors of production while the new economy adopts the so-called knowledge economy based on factors, including technical knowledge, creativity, intelligence and information.

United Nations estimates accounts 7% of world GDP. Thus, the knowledge economy must be the main engine of knowledge for economic growth. The knowledge economy is characterized by innovation, education, and infrastructure of information technology.

The main forces that oppose to the knowledge economy are working to change the rules of trade and national capacity in the knowledge economy, including:

• Globalization.
• The information revolution.

- Proliferation of communications network.

Knowledge economy plays an important role in the knowledge society, globalization that is dependent on the economy; it is also contributed to the result of knowledge in technology developments, according to the following requirements:

- Establishments of organizations and new economic rules.
- Opening up world markets.
- Redrawing the map of the world economics.
- The emergence of new centers that depend on world trade.

The emergence of the knowledge economy has led to emphasis on the importance of education as a key to economic success and knowledge society which is closely linked to the knowledge economy. All these are dependent on creative mind generated from higher education, which contributes a significant role in the production of knowledge. Therefore, there is a relation between universities and knowledge economy and knowledge economy relies on knowledge production and the production of knowledge is one of the most important functions of modern universities.

According to the logical perception of the importance of the knowledge economy and adoption of standards, the UNESCO institute for statistics adopts the following indicators:

- Total expenditure.
- Expenditure intensity.

Comparison between countries of the world in research and their potential revolution, each rising in industrialized nations, falls in consumer countries for industry, which confirms the link between arbitrator and higher education.

Within the prospects of the knowledge economy, the government may support "multiple channels of higher education, including research and development" and in the area of health and medicine, in particular the increase in life expectancy during the last century 1900-2000 about 30 years, consequently resulted in an increase in the imports of society about 2 to 4 trillion dollars.

In order to play the Iraqi university prominent role in the knowledge economy, learning process is necessary as economic activity contributes essential role in the production of knowledge which is necessary in a series of knowledge society, and that the Iraqi university should resolve many of the problems such as:

- The link between higher education and the labour market.
- Strengthening the IT infrastructure and knowledge.
- The integration of contemporary techniques in the processes of learning and teaching.
- Trying to remove the traditional features of these universities.

The world economy also reminds us, that it is moving towards a knowledge based economy and this trend is explained by economic theories, including new growth theory. This theory says that sustained growth (rather than growth for a short period), is directly dependent on three factors:

- Technological level.
- Technological growth rate.
- Saving ratio.

The traditional factors of growth, in which capital and labour are involved in directly in the growth equation. These changes - which are

very important in the world economy - are considered as added value that comes from the high technological level, the technological growth of the state and not only from capital investment and work forces.

Economic growth depends certainly on the researchers, discoveries and inventions and to those who invest these inventions and creations. This is in addition to the good implementation of these creations that depends on skilled workers, who can deal with modern means of production.

Many countries are no longer depending solely on industrial zones and free zones, because this mechanism does no longer believe in acceptance of added value. These countries began to adopt new patterns since eighties, such as, technological areas, science and technological cities and areas of knowledge. The objective basis for these patterns is the maximum utilization of new and innovative ideas which emanate from research centers and universities. The link between research and development on one hand and industrial and service activities on the other hand, is the nucleus of these areas or new cities.

The technological incubators are those of new patterns that were adopted for the purpose of achieving the objective mentioned above. Technological innovators are effectively helping to move the idea of a new form of laboratory or experimental or academic to the model proposed for breeding technique. The incubator creates the company and owns shares of the company's budget; the technological incubators are the best means of developing countries or developed countries to encourage the establishments of the initiators of their companies and are product of proven economic.

The technological incubators are of the entrances adopted at the global level, to encourage and support small and medium industries, where there are today more than 1500 incubators operating in the world focused in the USA, Europe, Japan, and there are 500 incubators in developing countries. There are about 200 incubators in France and more than 100 in Britain and 200 in Germany. Furthermore there is Japanese experiment in the field of incubators and science parks.

The communities in general must use the modern technological developments and the appropriate environment form. These communities, which would deepen the work of free thoughts, contributing to industry change. Among these methods are business incubators and small projects.

The business incubators is the integrated system of small projects that are, at the starting stage, in need of special support and protection so as to enable them later to move to foreign labour markets.

The overall objectives of the incubators are the development and creation of innovative projects, assisting owners of innovations and inventions, providing support and funding and providing services to destination funding for research and knowledge. One of the international experiences of such incubators is to implement the ambitious strategy for the developments of small projects; it requires a central body to manage, implement and follow up these strategies. The American experience is tracking U.S presidential administration but in France and Malaysia through specified ministry. In Egypt the incubators is represented at the social fund for development of the presidency of the council ministers.

There are currently in USA over six hundred of technological incubators such as Austin technological incubators, were used to reduce the failure rate for new projects, 50 projects have been graduated from the incubators, and 10900 new jobs were developed.

There are many types of incubators according to the objective for which they were established, including:

- Regional incubator
- International incubator
- Industrial incubator
- Specific incubator
- Technical incubator
- Research incubator
- Virtual incubator
- Internet incubator

The requirements for admission to incubators are place for the project, financial support, and technical support and skills developments. The incubator is preferably be located adjacent and not inside the campus of a university or research center in order to benefit from the resources and applied researches, laboratories and workshops, services and professors, but its presence within the university hinders the entry of customers because of security concerns.

Scientific incubators in the Arabic world

The incubators are working to accelerate economic growth by supporting the establishment of the main engine of the economy and its dynamic and high annual growth rate. The establishment and technology that produce goods with high added value requires an incubator for the care of these companies. These incubators provide opportunities to

accelerate the development of vocational skills for young people and specialists in the field of ICT incubator that was founded in Damascus between 2004 and 2006.

Technological Incubators in Egypt

The Social Development Fund adopted business incubators and technical mechanism to support the establishment of small projects and to develop the skills of self-employment among the initiators of technicians. Accordingly, the Egyptian Society for Small Projects introduced incubators in 1995 which approved the establishment of 30 incubators in Egypt. Then, it established 9 incubators that rely on simple technology in providing services, and light manufacturing projects which are dependent on knowledge and information such as the incubator of Mansoura, Asyout. The technical incubators are located near or inside the universities and centers of scientific research. The technological incubator of Mansoura University resembles the specialized informatics and biotechnology incubator in the Mubarak City of Alexandria. It is worth noting that one incubator accommodates about 40 projects to continue within incubator for 3 years. Statistics indicate that 520 associates will enjoy the services of incubators in 2006.

Science policies in Iraq

Science and Technology play a crucial role in shaping the challenges faced by individuals, organizations and nations constantly, (discoveries of genetic engineering, industrial human parties, mobile phone, biotechnology). The challenge facing mankind at the beginning of the twentieth century is how all countries could benefit from the strength of science and technology.

However, Iraq was unable clearly to improve the use of available science and technology, like other Arab countries, despite the availability of consulting firms and construction companies, millions of university graduates and about one million Arab engineers and hundreds of industrial companies and thousands of universities teachers.

The challenges facing Iraq, lies in the two groups, first resulting from major developmental problems, (food security, health, housing human rights, education, transport) and difficulty caused by the absence of the required scientific culture. The second is cultural in nature and include a special site independently. According to this scenario it requires the creation of systems of national science and technology take upon themselves the development of science and technology policies.

Iraq made over the past decades considerable progress in several areas, has increased the resources allocated to education, social services and infrastructure which has had a positive impact on the average per capita income and quality of life, followed by problems of concern (during the nineties). The per capita of the total GNP was decreased, the blockade halt efforts to diversify the economy and the adoption of the main sources of GNP on non- renewable mineral wealth.

There are some hypothesis for scientific and technological policies in the country which are limited in scope, effectiveness and ineffective strategies in the best of circumstances. So it was proposed to create new structures after some test on developed countries and developing countries.

The availability of financial resources in Iraq, especially before the nineties of the twentieth century has made great efforts in manufacturing, especially in the field of Military Industrialization. Iraq succeeded in building an independent military industrial base by enabling technology in modern manufacturing processes, however, strategies and manufacturing policies was clean from effective action to develp local capacity and providing appropriate incentives for local people to be able in modern industrial technology.

Despite this, Iraq still needs to pursue innovative methods to meet the daunting challenges of a large number of other sectors of production and services, as well as to develop scientific and technological capabilities of local traditional industries to modernize and to address a variety of social economic problems.

The scientific efficiency in Iraq itself has a capacity of innovative crucial role in facing challenges, in spite of standard conditions faced by the departure of many of them outside Iraq.

The limited scientific successes that have occurred in Iraq are the result of the efforts of some institutions of science and technology (Scientific Academy, the House of Alhikma and the institutions of higher education). On the other hand, important achievements were made in building the institutions mentioned, as well as in human resources development, but the institutions mentioned and others are still far from an enabling a distinguished role in development. The linkages and synergies between the scientific and technological institutions of governmental organization and the business world remain weak, despite the existence of some versions of contracting.

The spending on research and development in Iraq, at best, less than its counterpart in the Arab countries and more discouraging regarding the outputs of science and technology, so that scientific and technological publications in specialized areas and the number of patents granted to institution and individuals are much less than the average of the corresponding figures in Other developing countries.

257

As a result, the status of science and technology in Iraq needs a lot of attention. The input and output of scientific and technological point indicate the deficiencies in information networks, computers, advanced equipment, scientific research.

Scientific institutions

These include, institutions (universities, research centers universities, atomic energy, etc.), their shortcomings in Iraq are one of the reasons that led to the absence of a scientific and technological policies and the recognition of the limitations. The ineffectiveness of administrative practices and the existence of structural deficiencies are symbolic recognition of the need to develop capabilities in the next decade to operate in an environment different from recent years.

These institutions are currently working on:

- Production and dissemination of scientific research.
- Rating knowledge, covering a range of disciplines and areas of application.
- Training of some ranchers.

And, therefore it requires a complete reform of these institutions revitalization to support them, and identification of high-level priorities by increasing the competitiveness and environmental compatibility, creating an effective system of funding policies and linking them to industry and social and economic activities.

Scientific and technological cooperation with Arab countries, foreign states and international organizations

The current levels of scientific and technological cooperation between Iraq and Arab countries are very low, as well as with foreign countries which are almost non-existent as illustrated by the lack of joint scientific research and the outcome of joint publications. This refers to the need for Iraq in its quest to strengthen their scientific and technological interdependence.

To enhance communication and cooperation between Iraq with other Arab countries is a prerequisite in better determining the scientific problem and to obtain acceptable return of the available knowledge. Lounge of the basics of scientific cooperation is to avoid free turn key contains from any type of Technology, which is almost a priority involving scientists from two or more research fields, and incur the available possibilities to researchers to attend scientific meetings and make more use of international organizations.

The activity of science transfer

The term technology transfer indicates that the technology is acquired through the transfer of goods or services technology and therefore is not equipped for transportation and technological deficits.

The experience of Iraq in the transfer of technology is varied and appropriate in a large proportion, completed with foreign companies which provided comprehensive and complex technological deals in the framework of international market strategy, and the country suffered from indiscrimination of transfer that took place in the absence of a sound domestic policy in various technological fields.

There are two problems facing Iraq, the first is concerning the search, transmission, absorption, development and improvement of modern technology, and second is related to technology development. According to that, Iraq needs in the area of technology transfer the following:

- The search for technological alternatives.
- The selection of appropriate technology.
- Adapting the selected technology.
- Identifying the problems of adapting modern technology.

Research and development "R & D" activity

The research and development are both vital and active in maintaining the quality of scientific personnel, in ensuring access to advanced science and promoting technology transfer. In addition it is provide early warning preparation of technological progress, industrial, agricultural and health, whereas its investment is guaranteed.

The efforts in research and development in Iraq may differ from what made similar efforts in other countries in terms of outcome, it is still inadequate to meet the challenges posed by scientific and technological developments and the process of globalization despite the allocation of funds required in this area, and the last twenty years is good evidence.

- Establishment of an information society.
- Innovation of small and medium-sized enterprises participation.
- Development of human potential through the training of researchers.

Science policies in Iraq (Higher Education and other sectors)

There are no clear scientific and technology policies in Iraq, but may be the presence of strategies development that include the lines of long-term development of science and technology indirectly that deals independently within the development process. In Iraq private enterprises will invest in science and technology (Higher Education ………… and others) in establishing a scientific and technological infrastructure.

The scientific, international cooperation in Iraq is limited and weak at present, while the efficiencies gained by scientific institutions in Iraq in coordination with existing competencies in the areas of social economic activity exists in society.

Under scientific siege, budgets reduced drastically, and scientific institutions reduced also their expenses strongly which led to the non continuation of the previous level of innovative capacity, and the inability of the infrastructure and science and technology to do their core functions.

The formulation of science policies

Iraq continued to practice, so far away from any genuine interest, the development of science and technology policy, due to many reasons already dealt with. It is worth mentioning here that any scientific and technology policies, in particular, depend on the quality of Iraq's exports. Petroleum Exporting Countries and Iraq, supplier of natural non-renewable sources reflected the policies of science and accordingly, it requires the decision-making process and scientific identification of reliable channels between scientific community and the political leadership. Moreover, the wording of policy drafting requires determining the objective of science and technology that includes:

• Promotion the growth of local companies specializing in services and manufacturing activities.

• Advancement of specialized institution.

• Increasing the movement of personnel in science and technology.

• Overlapping science policy and technology with a range of social and other economic activities.

• Development of institutions that will depend on science and technology that operates in a different environment.

• Strengthening the management structure of science and technology.

• Raising the level of resources.

• The effectiveness of R & D institutes.

• Development of technology transfer.

• Development of international cooperation.

• Improvement of researchers' wages and working conditions.

• The development of science and technology in long- term visions.

• The establishment needed in long term visions of all relevant institutions.

• Definition of the roles of effective government, private sector institutions, non- governmental organizations and professional associations.

• Adoption of progressive methods to build scientific capacity.

• Removal of duplication, overlapping and conflict in science organizations.

• Monitoring and evaluation and effective drafting in the relevant agencies in science and technology policies

It is proposed that a variety of science organizations should be related to the formulation of science and technology policies, including scientific societies and other professional such as industrial and technological organizations.

These organizations have to operate in an environment different from the classical environments in recent years (currently confined to the production, dissemination and application of knowledge covering a range of disciplines and areas of application), but must evolve and interfere with a range of social and economic activities and cooperate with other private sector organization. Moreover, these organizations propose concepts and scientific policy inputs.

Moreover, these scientific organization should involve the largest possible number of workers participates in the preparation of science and technology policies, including representatives of government departments concerned economists, chambers of agriculture, industry, trade, non-governmental organizations and scientific societies.

Institutions of higher education

The performance of institutions of higher education in Iraq is weaker than the performance of the countries that have the same resources and level of development. The reasons are due to organizational structure, prospects and understanding science policy.

It is noteworthy that all efforts made in the past to link universities with proposed scientific strategies were not successful for the same reasons. Therefore, it requires the involvement of these institutions in the channels that suggest implicitly of the scientific policy and development new scientific boards to discuss issues of science policy.

The efforts in research and development in Iraq may differ from what efforts made in other countries in terms of outcome and the challenges posed by scientific and technological developments and globalization. The linkages between research and development do not meet the need, as they are usually not based on cooperation, and are sometimes contradictory because of the lack of flexibility and bureaucratic practices.

The problems on research and development requires from its institutions to evolve, according to the ideas that have been developed. These relate to the maintenance of the quality of special scientific cadre. Further more then to access to renewable science in the world, promoting industrial development, technology transfer, contributing to social, and economic planning and accordingly proposed that the research and development institutions have special functions as well as:

- Proposals should be submitted through the channels when developing the new science policy.
- The possibility of providing early warning systems in preparation for preparation for technological change.

It is proposed in this area, that the scientific academy have a complex role in the development of science policy and is considered to be one of the main tributary of the scientific policy and may be a substitute for the supreme for Science Technology. The participation of specialized committees for example, could provide complex scientific projects, relating to environmental pollution, energy and natural resources projects and areas of agricultural development, public health and natural resources.

The specialized scientific societies could propose several projects and comprehensive survey of scientific resources, human and various national materials. Also enhance the working groups that can be developed and networks of cooperation between scientists from various disciplines concerned, with the establishment of networks of information for selected sectors based on the priorities identified by policy. It can evaluate research projects and public awareness of science and technology and increased attention to scientific research.

These institutions are required to have the presence of a specific set of targets linked with science policy and technology, which are consistent with a coherent vision for the future.

The responsibilities to the views of specialized advisory functions include characterization and strengthening the linkages and knowledge flows and providing technical services.

Proposal to develop a higher council for science

Iraq's various institutions face difficulties in promoting scientific and technological capabilities and future planning of the scientific, issues to acquire the capacity of innovation. This can not be achieved unless being done within the framework of science policy interrelated. Accordingly, it is required to use new methods of science and technology policies, and to develop integrated growth strategies that take into account of local and external scientific conditions. To implement these, it is proposed to develop an academy or a higher council for science and technology. This Council (or academy) which is an institution with personal, moral, financial and administrative requirements characterized by autonomy should be linked to the Office of the Presidency.

The council carries out the followings:

- Development of science and technology policies in accordance with the demands of current and future science.
- Contributing to the scientific development of internal and external sources.
- Promotion of scientific studies and research in the country to keep abreast of scientific advances in the world.
- Establishment of scientific ties and close cooperation with Arab and international destinations.

The academy or the Supreme Council consists of:
- Members of not less than 35 not more than 40 including the President of the Council or the Academy.
- Secretary-General.

The member should be a scientist and researcher in one of the branches of knowledge (agricultural, industrial, treatments, medical, engineering) and has a broad access to one or another branch of knowledge, and has genuine scientific product.

- The academy members should be appointed by the Prime Minister and enjoys the rank of Deputy Minister.

- The Academy full-time secretary-general is appointed from among its members.

- The Academy has a number multiple specialized committees working in coordination with other scientific institutions in the country (Commission on Technology, the Committee of Water, Energy Commission and Commission for information).

- The Academy has specialized offices within the framework of scientific knowledge (agricultural, medical, engineering, pure science).

- The Academy creates the suggestion of science and technology policies and then filed for the presidency for the Adoption (after being revised, by its committees and services).

- It also works with the recommendation to grant material assistance to the centers and individuals, adoption of the establishment of various scientific centers.

- Development of labor contexts between scientific institutions and the beneficiaries and the mechanism of cooperation between scientific institutions of Iraqi and international institutions involved.

The Supreme Council for Science and Technology proposes a policy for science and technology in the country in accordance with the following tasks:

- Coordination of the efforts to formulate science and technology policies in Iraq at the institutional level.

- Clarification of methodologies used in policy formulation of science and technology.

- Modernization and integration of science policy with development policy.

- Analysis of scientific and technological capabilities in Iraq with a

focus on research and development.

- The need to inform policy makers of the methods used in the planning and programming of research activities

- Building local capacity through education and experimentation.

- Following up the implementation of these policies through the secretariat and a mechanism to provide on issues.

Formation of a Higher Council for Science and Technology

Detailed proposal:

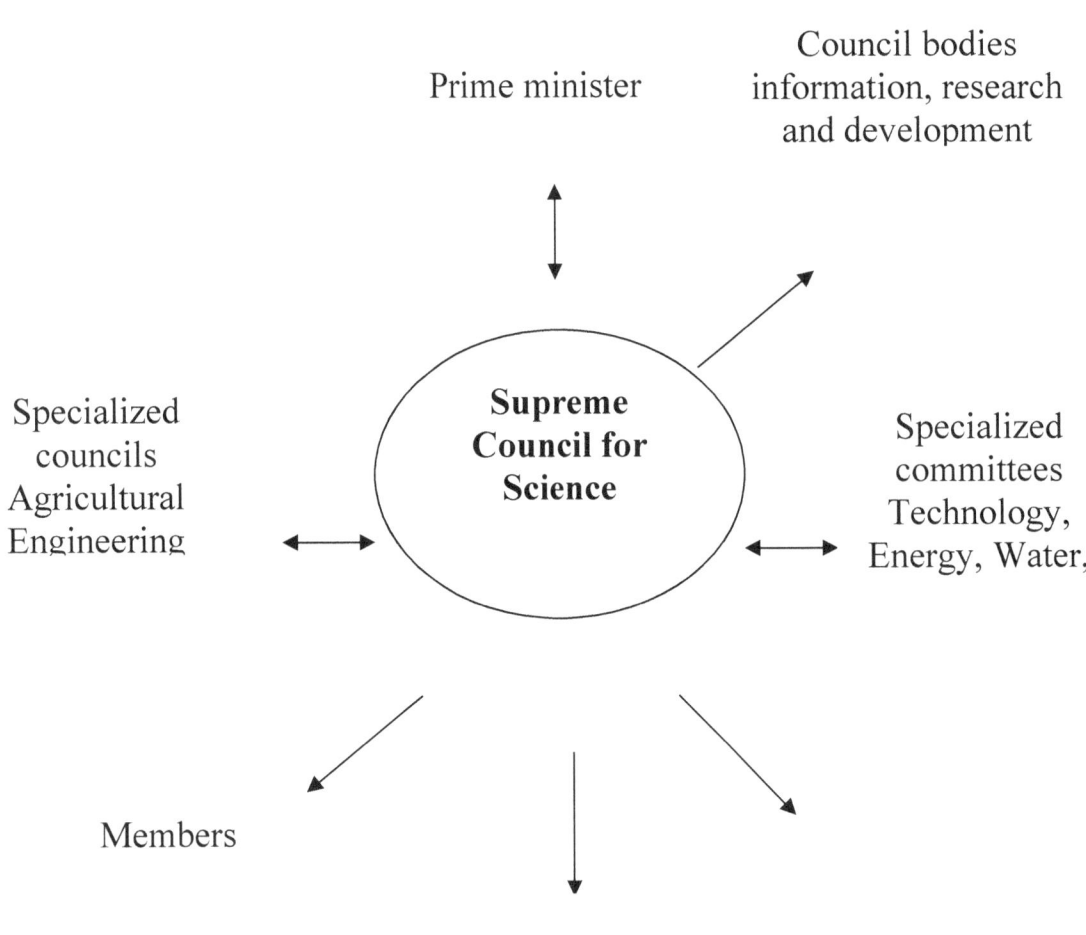

Prime minister

Council bodies information, research and development

Specialized councils Agricultural Engineering

Supreme Council for Science

Specialized committees Technology, Energy, Water,

Members

-

- other countries, with balance between the two sides.
- Investment of scientific institutions of science and technology in creating the scientific and technological infrastructure.
- International scientific cooperation.
- Promotion of development of national technological capabilities in both the   private   and   public   alike.

Coordination of competencies gained by scientific and technological system and their integration with national huge number of competencies in the areas of social and economic activity.

- The integration of science and technology policies with a set of national policies and programmers that meet this purpose.

- Continuation of production and dissemination of scientific and technological knowledge and maintain an effective level of innovative capacity.

- Continuous documentation of formulating policies and scientific technology.

-

General trends

Policies can be characterized by scientific and technological trends and are generally maintaining the balance between any movement toward the development of local scientific and technology and the establishment of links with external sources of technological knowledge that take place in several stages:

• Continuous assessment of the current status of science and technology.

• Development of the prospects of continuing to develop science and technology during any (year).

• Development of a strategy based on specialized studies related to the objectives of developmental policy.

• The preparation of a detailed five-year plan accompanying executive programs constitute the followings:

- Keeping pace with scientific and technological developments in the world (biotechnology, information technology, energy technology… etc).

- Keeping up with scientific and technological developments in the world through the follow-up scientific and technological capabilities such as (biotechnology, information and energy).

- Monitoring the change in scientific institutions and the Arab world and the extent of the involvement of scientists in Iraq with their peers in the Arab countries and the world.

- Following the changes in social and economic conditions throughout the world.

- Examination of human skills constantly has been much more important than raw materials and it leads to many possibilities, including:

- Achieving parity with other countries in science and technology.

- Identifying better scientific problems and understanding of the issues that are raised in a more comprehensive way.

Other organizational matters related

These include the input and output of science and technology policy with emphasis on the role of different institutions, including industrial, educational and private and scientific academy. It is also important to emphasize in this area on organizational matters that are between institution and how to submit proposals and receive guidance on science policy that includes:

- Providing specialists for ongoing assessment of the potential of science and technology.

- Involving the largest possible number of parties involved in the preparation of science policy as well as interaction with the production sectors and services through formal mechanisms.

- The participation of the private sector in financing scientific capacity building.

- Consideration of responsiveness to the requirements of development policies.

After gathering proposals for science and technology policy from various channels according to directions from the supreme authority, it is proposed, in this case, to adopt scientific policy objectives (competition, trade barriers, the environment, new technologies). Then the policy with the top destinations should be approved by political leadership.

Thus, according to this scenario it is needed to start work for the establishment and completion of institutional arrangements and accordingly the following are accompanied the proposal.

- Evaluation and forecast of scientific and technological advances.
- Assess the market demand for science and technology.
- Planning and management of science and technology.

The scientific policy decision should take into consideration the following:

- Be based on local capacity and scientific progress in other places and to balance them.
- Contribution to regional and international programmes for the development of science and technology.

The role of the Supreme authority:

- It is proposed to take measures immediately to develop a policy and scientific and technological integration in social and economic development plans in Iraq, and the establishment of linkages with the private sector through specific proposals.
- Creation of private channels and specialized committees to prepare the requirement of science and technology policy and guide these channels with the supreme Council or academic parameters of this policy.

Financing plans and activities of science and technology

It is required at this time highlighting the role of financial institutions and the mobilization of resources to allocate sufficient resources, to invest in technologies that have been obtained or that have been developed locally. By financial mechanisms it is necessary to implement policies, by increasing allocation of resources for scientific and technological activities, research and development budget in public and private sectors.

First: The state budget (current and investment).
Second: The socialist sector companies.
Third: The mixed sector companies and private sectors.

In order to establish a national fund for science and technology it is required to have feasibility study, allocated 0.25 per cent of Iraq's national budget, to finance projects of technological priority implemented through contracts awarded on the basis of competitive bidding.

The funding system in the private sector, which relies on the so-called policy of risk sharing by adopting the industrialized countries, ranging from tax incentives, grants and loans to encourage private investment in innovative activates

Chapter six

Sciences the first in Technology

Preface

Academic disciplines have evolved with the development of sciences and various new disciplines such as engineering, agriculture, science and total treatments that began in the nineteenth century, whereas the twentieth century indicates other disciplines such as business management, journalism, information and library science, economics, politics and world affairs were added. Each state has its own special methods to determine its own disciplinary university and identification numbers, graduate students and the quality.

The world witnessed in the twentieth century breakthrough in all fields and scientific trends, so there are no boundaries between different disciplines. For example, medical science requires engineering science and recent tests of modern science depends on the physical, chemical extraction and analysis and also relies on mathematics to lay the groundwork mathematics.

The progress and development in pure science, for example, develop new subjects and disciplines and specialties of science. New interfaces were not known during the first half of the last century. The results of these major changes in curriculum and build up research transformed these developments to the university curricula. Seminars and researches are now carried out in different ways including:

- Bachelor based on the study and theses.
- Some universities in Britain and Germany developed curricula at the level of initial studies that include research and study.
- Divide the present fields such as industrial chemistry, chemistry of life with medical side and other disciplines in the branches of pure science.
- Develop competencies.

These terms of reference have been developed in American universities in physics, chemistry and mathematics, to prepare graduate in some sectors such as engineering, chemistry physics and chemistry, agricultural engineering, and medical studies.

- Adding assistance topics.

Some topics have been added as assistance of many of the terms of reference of pure sciences, including education, literature and library services and use modern machinery and computers.

- Other disciplines (Sandwich)

British universities were carrying out by expanding the initial years of university study for use in increasing opportunities for the systematic teaching and applied for and rehabilitation work in various production sector.

Bio-chemistry

Amazing developments have taken place in the chemical sciences particularly during the second half of the century, including implicit and other interfaces. Developments on the implicit content and the vocabulary and mechanisms are known in chemistry and provide improved or new interpretation of events and phenomena and chemical reactions, as a result also of new subjects and disciplines within the science of chemistry itself. These developments have led to the opening of new channels in scientific research and technological innovations such as chemical industries to create new chemicals, or chemical industries, and new techniques.

The developments of the second type of chemical sciences interface had addressed the disciplines of science linking chemical sciences and applied various treatments. These developments have led to the developments of science or the new terms of reference were not known before.

Bio-chemistry studies the chemical and physical characteristics of the components of the cell and features of the life systems of the components, as well as the interpretation of what these systems in the cell Biochemistry provided a lot of accomplishments, it has helped to clarify the mechanism of medicine and contributed to the diagnosis and treatment of many diseases and provided techniques which could be used measure the level of many of the compounds in vivo.

Biochemistry lasted over the age of a century in different disciplines, some with a study of the materials that make up plant cell and then called the chemistry of plant life, and then which is related the animal cell which is called chemistry of animal life if the human cell is the target.

Chemistry has expanded to clinical biochemistry that includes chemistry of life, becoming a physical, organic and biochemistry and inorganic chemistry as well as nutrition. Interested in chemistry, life functions of the modern systems of life, have contributed to the means of study in the last century with the observation of these systems directly during the work, either at the present time which has changed the picture and it became possible to obtain the most desirable observations by the development of viable technologies (electron microscope, radioactive isotopes, Immunology, spectrum).

The scientists believed at the end of the nineteenth century that it is possible to obtain some information relating to the systems of life, by studying the chemistry of cells and for decades was followed by chemists adopted the chemical methods available and succeeded in obtaining useful developments. Significant improvements to the technical methods such as the use of chemical isotopes have greatly increased the sensitivity of diagnosis of different types of molecules of life and others, and when it is necessary to separate the components of the chemical reaction through life and is very sensitive, then used deportation electric traditionally.

When the attention has turned physicists, chemists, physicists about the science of life (and perhaps due to the ability of living cells to configure the system, although the laws of physics, emphasizes the universe there is a tendency towards non-attendance) then emerged the technical methods of physical, chemical, physical, such as spectroscopy, diffraction to be applied an the field of biology.

The progress achieved in the chemistry of life has begun to acknowledge that the livelihood systems containing small particles interested m organic chemistry to study and clarify as well as large molecules called macroscopic particles which are not molecular weights less than 100 million times the mass of one atom of hydrogen. The importance of macroscopic particles of the life systems in its ability to privacy in life interactions composition of building blocks, and can say clearly that he had made in the past years considerable effort to characterize the annexation of macroscopic particles as well as the reactors that occur between them and the need for advanced methods of separation and purification and characterization of macroscopic particles in order to obtain information on structural composition of the macroscopic molecule.

The objective of biochemistry for nearly half a century is to collect and organize interactions that occur in living cells. The motivation for this major effort is that a significant number of the attributes of living cells can be understood through these interactions that are typically characterized by the formation or breaking covalent bonds. It is been clarified on the liberalization of energy as a result of break chemical transformation processes as well as molecules of life and mutual assembly operations amino acids, sugars and fats to form macroscopic particles.

During the last thirty years clearly demonstrated that the reactors that occur between molecules due to physical, those that are not or break covalent bonds have the same importance of chemical reactions, for example, that the organization of chemical reactions (i.e., the degree of permitted them to occur) performed by the physical changes that occur in the structure of (construction) of large molecules, as well as the creation of active centers in these molecules and the resulting interdependence of the non-covalent small molecules, in addition to, many of the qualities of a macroscopic aggregates molecules in cells or in the organism (the cell membrane and walls of cells and chromosomes).

Plurality of molecules of life structure consists of installation of the first structural molecular structures of multiple different types of units place (serial), for example, the sequence of amino acids found in proteins and sequence by chemical analysis. The secondary structural composition which involves the formation of a complex three-dimensional structures is called to direct all of the units for multi-particles to other units and is called the secondary structural composition tradition or (body and image) or the status of the foundation structure or backbone of multiple chains. The forms, which consist of surfaces and different types of these mixed forms, and called on the direction of (position) of side chains relative (amino acids, nucleic acids or bases) triangular structural composition. A lot of multiple molecules of life with each other to be as complex as the structures of several structural units viruses, membranes and capillaries bonds and are usually in one level, where you specify the types of bilateral structures of proteins. On the other hand that includes the alpha carbon to allow for many types of structural combinations. The two phosphate ester bonds in nucleic acids are subject to sag as well, because the flexible rule and hate water and one level surrounded by a few of so they are usually located one above the other, thus reducing the adhesion of water, and this increases the structural rigidity of installation.

The multi-life linear molecules, which has no free rotation about the bonds, which do not interact aggregates side is called the file is not a random combination structurally specific dimensions or size of distinct wraps by Brownian movement. Size can be measured by the value equal to the rate of rotation of the radius around a point or an axis.

A nucleic acid represent the brain of the cell brain cell with a specific developed program, to be issued through the instructions for that cell fusion and installation of the life and death and plan for the future.

There are two types of nucleic acids (DNA, RNA) both of their differences centered a long chain molecules composed of nucleotides and position of certain forms.

In 1953 Crick and Watson was able, who have previously received the Nobel Prize in 1962 developed a model for the DNA structure, consisting of two strands of units of the four nucleotides arranged in orderly fashion, and every one of them is a multi-helical nucleotides wrapped around a common axis to form the double helix right direction.

The models explained by x-ray, have two sections through the longitudinal axis of the first of 0,34 nm and the second 3,4 nm.

Model of "Watson & Crick Model" Both Watson and Kirk in 1953 the first specialist in genetics and the second physicist to develop a model structure that represents the structural basis of DNA. Jarkav in the light of studies on the percentage of nitrogen bases, X-ray dimensions, as well as the fact that adenine = thymine and Guanine = Cytosine

Synthesis and structural forms of DNA

Features of the model of Watson and Crick, this is called beta-form and include:

- The DNA is composed of two strands of dioxin nucleotides wrapped through multi-spiral system.
- The nucleotide chains are connected by diester bonds within one strand.
- The bases purines and pyrimidines facing each other, so that in particular, adenine faces thymine and guanine faces cytosine through hydrogen bonds.
- The order of nitrogenous bases one in series vary from other.
- The levels of sugar rings parallel to the axis of and phosphate groups abroad.
- DNA is divide in two parts the first is called water hating (nitrogenous bases), located internally and a second which is externally faced of the surrounding water molecules containing phosphate groups.

The second half of the nineteenth century witnessed a series of discoveries of life such as, a serious (cell) theory at the hands by Matthias in the plant and then Theodorishvan in the animal. Since of plant and animal are composed of cells, that evolution (cell theory) and their development is considering critical stage in the progress of life science similar to atomic energy The human Physiology science as one of the branches of the life science that refers to the amazing facts explaining the greatness of the Creator and accuracy of the details and secrets. The digestive system for example (the greatest chemical plant in the world) including, the methods of food analysis of chemical analysis of various surprising and distribution of fairly Safe food distributed to millions of living cells. In view of these living cells the issue of causal efficacy and secret of life that fills justify the astonishment and admiration for self-cell, while adapting to the requirements of their position and circumstances.

If we explore the science of life, we will find another secret of that biggest secrets, the secret of the mysterious life, which fills the moral conscience of mankind, with the concept of divine fear and faith, firmly established in it. The theory of self-regeneration was collapsed at the depth but the unequivocal scientific experiments, demonstrated the invalidity of the theory of self-regeneration. The material basis of life science was examined and then basically spread the idea of elements. The atoms are spread better for the basic materials of the universe and second nature that the elements consist of a central core electrons of the nucleus orbit (negative) and the nucleus contains protons and neutrons. Attempts were made to alter the material to absolute energy, no electric charge. In other words removing character from element in the light of the theory of relativity of Einstein, where the body mass is relative, not fixed, and increase with the speed according to Einstein equation energy = mass of × square of the speed of light and mass = energy ÷ square of the speed of light. As a result, the atom, including of protons and electrons are condensed energy. Appeared in various forms and multiple images, whereas materials has been converted into energy and energy to the material.

It follows from the views put forward that the original materials the world-life the reality show one common in various forms, and the physical properties of compounds are accidental such as the liquidity of water is incidental, not self-evident, since it is consisted of two toms and possible separation these two elements from each other and the status of water disappear completely.

The characteristics of the simple elements themselves are not self-rule but are incidental to the material. That such material characteristics become the light of the above facts incidental, it is encroached to be among the identified energy and philosophically, the presumption of the material in the world of life on the top reason capable for denial, as well as of effectiveness.

Genetic Engineering and Biotechnology

Genetic engineering caused major developments in life sciences, including applications in medicine, which includes diagnosis and treatment.

The concept of the genetic engineering and technologies based on multi-splitting of DNA By special enzymes work on specific sites and then linking the pieces formed with DNA from other sources, is then the proliferation of hybrid is able to reproduce on vehicle "Cloning", including bacteria and viruses which have been used to develop genetic engineering techniques and other such as electrophoresis and auto radiography.

This technology can be used for the preparation of special DNA sensors for the purpose of searching for specific genes or specific parts of the DNA technology to clone parts of the DNA in large quantities by the use of the enzyme "PCR".

Applications of genetic engineering:
- Gene therapy: It is used in the treatment of some diseases, including the treatment of brain tumors and in the reduction of cholesterol in the blood.
- Genetic mapping in humans: one can look at the matter in the future after the location of many human genes.
- Stimulation the immune system to produce antibodies more efficient and accurate (a vaccine against the virus of hepatitis "B").
- Production of proteins of medical importance.
- Determination of the nature and location of genes for some genetic diseases.
- Diagnosis of genetic diseases before birth.

Biotechnology caused enormous developments human, including those which were aimed at human use:
- Industrial electronics.
- Products of biotechnology space.
- Environmental treatment
- Extracting of oil by microorganism

- Medicinal plants.

The transfer of chromosomes from one cell to another is of chemical and living concept for the process of cloning, and its backbone. The genetic chromosomal transfer is not considered new, it has been exercised during the implantation of an egg in the uterus of a female in (1978) (the birth of a tube child - Louise).

Specifications and features of the cloning by the cloning is characterized with specific characteristics of chemical features of some new and some old concepts. These specifications are as follows:

- Converting an adult cell (totally grown) to a cell could reproduce without vaccination.

- Converting the reproducing cell to full a creature breeders (replica of the mother).

- Transfer of mature cell grown to fully grown live immature egg in the uterus of another object (the sheep, for example).

- Development of a new principle called the principle of Wilmurt can be expressed as follows: (very specialized mature cell from an animal that awakens static sophisticated genetic information in the chromosome to become a source of a whole new creature).

- This process of reproduction is not a sexual in the traditional sense (without sexual contact between male and female).

- The genetic cell cannot be used through challenge and breeding.

- The cloning from adult stem cells facilitate the researcher to wait to see the nature of the thing itself before proceeding to reproduction.

Features applied to clones (the experiment of Ian's death)

- Converting a cell of the nipple of the gland of sheep (full growth-extremely) to a cell with a capacity of breeding and without sexual relationship.

- Transfer the mature full cell to fully live the immature egg in the uterus of a sheep.

- The cloned sheep of inherited properties, from the donor mother (the birth of Dolly the sheep).

- The sheep (Dolly) is not the nascent daughter and her mother, can be its twin.

- The sheep of the cell prepared for a clone of the breast, containing all the genes necessary for a complete sheep.

cloning

The cloning technique, distributed according to principles and specifications on several areas, including gene cloning embryonic cloning and cloning by nucleus. Gene cloning refers to the production of similar genes similar to the original gene, and the best example of this, is the process that occurs naturally when giving birth to identical twins, as a result of one split of egg genes and its distribution into two cells identical to the growth of each to produce transferred uniform embryo. But when the separation of embryo's cells from each other then process is referred to by certain embryonic cloning, each cell is growing separately to produce an integrated organism. Researchers have successfully cloned monkeys and frogs from embryonic stem cells match similar to the original. It was easier to deal with embryonic stem cells because they do not have any discrimination that they were not yet have been developed to convert to and brain, muscle and other organs, in the wombs of their mothers. But it turned out with the progress of development, the property of discrimination occur a change in the DNA. The passivity of embryonic cloning researcher stresses the intrinsic character of being unaware of what will emerge.

The third cloning, which represents the third occurrence of mature cells, a researcher can wait to see the nature of the thing with his own eyes before they proceed on reproduction. This represents a new type of reproduction, which is using the nucleus of an adult cell, adopted by Wilmot the specialist knowledge of embryos, which represents technical innovation for the reproduction of fetal sheep from an adult cell, and he was able to clone sheep, genetically modified by human genes to produce factors that deal with clot the blood. The Dolly, of Wilmot considered as the first which was cloned with the latest technology and adult cell cloning technology has international new principles of life, especially relying on a somatic cell, and successful steps can be carried out each and every one has a recipe associated with these principles:

- Dealing with the cell donor (host) intended for cloning and extracted from the membrane cells of a pregnant sheep is characterized by white color and necessary genes needed to form the purpose complete sheep. This step represents a use an adult cell transferred to a cell with the ability to reproduce without sexual mating to become a creature without full sexual intercourse and instead of being subject to embryonic cell fusion.

- An unfertilized egg cell that has been stripped from another type of sheep with a black color in the laboratory and the nucleus was removed, but the cytoplasm remain in the proper position.

- The defined cell donors exposed to famine and by halting its development and to prevent divisions and food resources for a period of ten days and surrendered to a state of sleep. The egg were put near the nucleus of the cell donor and prepared for cloning. Using weak electric firing bursts then nucleus and the egg unite and start acting as fertilized egg.

- Allowing the embryonic cells to grow and divide and then formation of embryo in the pot laboratory, implants in the uterus of sheep with black head. The sheep that was born become and was named an international replica of the white sheep of the cell donor not the same as the black sheep used for the lap.

The secret of life

When studying living organisms as secret of life and mysteries dependence of the of life it might be important to consider many concepts such as the idealism that reflect the objects of which they are the realities of life and is found independently of sense perception, and is the way of our thinking for our perception. As a matter of realism philosophy, another question is the position of living objects reason for all the phenomena of existence and the universe, or bypassed to another reason represented the deepest area of the material, another beyond the spiritual and the last living as a reason to click the Spirit a realistic concept of Divine (the divine), and this concept does not mean dispensing with the reasons natural or something to rebel against the facts of sound science, a notion that God is deeper reason for science to explore the wider area, including the continuing nature of living organisms. The Philosopher, whether materialistic or divinely believes the positive side of science, such as exploring the unity of life of organisms with the general knowledge that is not in issue, the scientific issues , divine philosophy materialistic philosopher material. Agasiz introduced in the year (1858) the idea represented by all kinds of organisms created by special acts of erecting force. This opinion is consistent with view of both Al-Razi and Pasteur and settled their minds with that every living being that must be generated by an organism like him. Also, Hermann Erhard Brichter has stated that each living thing is eternal and produce only from the cell.

The cell as secret of life

The human digestive system which is an arm of the human physiology of man, explain grandeur of the Creator and the accuracy in detail of the multiple and various secrets. The digestive system is sophisticated chemical laboratory having different methods of food analysis then the food will be distributed equitably to millions of cells that make up the human body consideration involving the secret of life and admiration for the cell. These cells are different technologies for tissue engineering and in the digestive system nearly two hundred thousand reaction within 24 hours. Some of which are the heart muscle shrinks and flattens millions of times during the whole year tirelessly, to obtain the necessary energy for thinking and movement and speech, including also the disposal of waste and toxins in the body, looking at the cell that its approach is one of the secrets of life. These secrets are adopted according to serious lay cell; it is one of the discoveries, theory at the hands of Haydn and Schwan then cell theory, considered a critical stage in the progress of life science, similar to atomic theory in chemistry. The cell in the body of an organism is also similar to personnel in the communities or the living cell act as technique specific to perform a particular job. The cell becomes a plant or accurate chemical plant. The nerve cell act as a system, for example electrochemical transformation of chemical energy to electrical energy and electrical energy to mechanical energy or kinetic energy. There are also some cells manufacture of the of hormones and other life products used as system used defensive attack their products all exotic and cells in the process of purification and filtration and the cells that absorbs. Furthermore arises from a single fertilized cell, tissues are various heterogeneous tissues, the different organs and different functions, the bones, muscles, cartilage and twigs, leather, and the blood vessels. Then a living cell had made specific technologies in particular, for example, nerve cell is electrochemical system that can shift the chemical energy into electrical energy and the last to mechanical or dynamic or may become a cell laboratory or chemical plant carried hundreds of chemical processes complex and there are cells specialized in the manufacture of hormones and the other to produce biological weapons of and cells of the nomination and purification.

It is clear that the tissues that originate from a single fertilized cell then divided into thousands of cells to materialize into the bones and muscles, cartilage, twigs, leather, and the blood vessels, these tissues are formed in the early embryos and mutate into organs and systems in a stand- alone, but integrated in the performance of its functions.

According to the information of originated from the cell that there is strange power lies in living cells. Walker, professor of Plant Physiology say (that components of a cell arranged in a strange way in which life emerged). But researchers still are unable to make blood cell components and accurate knowledge of this so-called the mystery of life.

Nucleus and secret of life

There are at the center of the cell mass of material in the spherical or oval clusters of objects in the body and continues to represent the mysteries of life and plans, regulations, and ideas of life.

In the cell of the human body forty-six chromosomes except the egg (cell reproduction female) sperm and egg, each containing 23 chromosomes (half the number in the human cell non-reproductive). The secret of life is due to this process of somatic cell fusion where each chromosome separated into two parts, it becomes in each of the cells of the fission 46 chromosomes. The chromosome (multi-genes), each genetic factor arranged in two strands one received from the mother and the other from the father.

The nucleus synthesize nucleic acids, seen in the central nucleus filamentous structures and spread over its surface of granules of quick-impact dyes include the nuclear network and the network should be clear when they are not in the case of splitting and dividing at smaller and thickening of these lines is called chromosome. The chromosomes in a cell division looks like similar pairs of fixed shape and fixed number for one type of living organisms.

(Russell and Wallace) says that the cell nucleus is not chemical but structure if analyzed and during the processes of analysis of the most secret mysteries of life may be lost.

Molecules of life, which are building the organism

The space of cell is containing the liquid water containing the various ions and compounds with molecular weights of small, medium and macroscopic, and it is possible to measure the ion composition in each cellular organelle, where each one of them has different ionic compositions.

The sodium ion "Na +" ion is the main ion out the cell in which the 140 mM / L is also found a positive ion in the fluid cell in the Interior position that the , is potassium "K +" Cation cell procedure. There is magnesium ion "2 + Mg" in all cellular spaces inside and outside but with lower concentrations of sodium, potassium and chloride is "CL-" ,the main negative ion outside the cell, with hydrogen carbonate ions "and" small amounts of phosphate and sulfate, and the proteins carry a negative charge at pH 7,4 in the tissue fluids.

All living cells contain a different chemical components of water 70-90% and 2-5% of inorganic ions such as sodium, potassium, chloride and sulfate, and magnesium ,carbonate molecules of life, as well as small, medium and macroscopic molecule that constitute 8-25%.

It has been proven that all the elements in the periodic table of Mendeleev's constitute in the composition of the organism divided into small elements and large. The carbon, oxygen, hydrogen and nitrogen constitute 96% of the elements in the cell, while calcium and phosphorus constitute 3% and each of potassium and sulfur, iron, sodium and chlorine 1% There are very small quantities of the elements iodine, magnesium, copper, manganese, cobalt, boron, zinc, fluorine, selenium and molydenom.

The chemical side is concentrated in the molecules of life on the carbon, which constitutes about 50% by weight bio- molecules are characterized by life-covalent bonds, four of which related to carbon stubs and have different angles of particular value from one carbon atom to another in different molecules of life and because of that there are different types of building structures with three-dimensional, these structures contribute to clarify the complexity of the cellular composition with particular reference to its failure, as well as various forms. In addition organic compounds are characterized by free rotation. The tetrahedral bonds emphasizes on individual carbon atom of the very important properties of organic molecules and the presence of four different groups or different atoms connected to carbon and the last become non-symmetrical (a carbon-atom covalently bonded with four different groups) and composed (which every one of which is mirror images of each other) with a symmetric arrangement in space and called isomers mirror of light for the chemical similarity of the interactions but differ in physical properties of the rotation of polarized light.

Mathematics and computors

Emphasis on the unity of mathematics and philosophy, includes several directions, such as, applied mathematics, engineering, life science, mathematical economy, financial mathematics, probability and statistics, operations research, mathematical research and other applications have also been reaffirmed through mathematics public programs, In particular, special importance to the modeling of mathematics, has gone some universities to develop a draft "project" and mathematical clinic confirm this applications. Moreover, studies have developed interface such as mathematics statistics, mathematics philosophy, mathematics and the economy, mathematics and physics, mathematics and engineering, mathematics and computers, and others.

Computers have been introduced as a mean to teach some subjects of mathematics and devise new methods of teaching an alternative to the currently prevailing lecture and attention to the employment of culture and mathematical awareness.

The unity of mathematics as a philosophy and the name of mathematical sciences include many aspects such as applied mathematics (engineering) and life sciences, mathematical economy, financial mathematics, probability and statistics, operations research, mathematics research, and others. It was also stressed the applications of mathematics through public programs, and in particular to give special significance to the mathematical modeling, has gone some universities to develop a draft "Project" clinics "Mathematical Clinic". Moreover, studies have developed an interface such as mathematics and statistics, mathematics and philosophy, mathematics and economics, mathematics and physics, mathematics and engineering, mathematics and computers, and others. Computers were introduced as a means to teach some math and attention to promotion of culture and awareness of mathematical mathematician. To provide students with the qualifications can be expressed as follows:

- A sufficient culture to qualify them to understand the difficulties.
- Encouragement self-confidence of all to address the difficult technical issues.
- The amount of uncertainty makes them ask the right questions.
- The amount of persistence makes them continue to search for appropriate responses.
- Discretion to choose what is right.

Informatics

Recent trends suggest that the evolution of Computer Science from the computer discovery and manufacturing in the early sixties. Computer science was not then separated itself along the lines of pure or applied science, but they grew up in the embrace of scientific departments, especially departments of mathematics and electrical engineering departments. The computer hardware and curriculum were not possible to sufficiently and necessary for the different compartments and focused the first two- way computer materials is the direction of engineering ties to computer components and architecture, and the second trend is the compiler. Number of programming languages in science and engineering faculties used to solve mathematical problems and mathematical, engineering or for the purposes of modeling and simulation. And the evolution of computers in terms of physical hardware "Hardware" and software operational "Software" to develop computer science as one of the basic science, where it was subjected to the axioms of this science and the philosophy of pure science and applied at the same time. Since the late sixties, introduced independent scientific sections of computer science "Computer Science" or computer science and information "Computer and Information Science" or computer education "Computer Education" or "Cybernetics" or processing of information "Data Processing", or under other refers to one type of information processing.

Development of computer education even intervened materials with a large number of scientific and humanitarian and computer hardware has become part of the curriculum requirements of the various articles of pure science, applied science and humanities. As a result of the marriage of computer science with other fields of science and new scientific knowledge, especially in the application, example such as field of computer linguistics, "Computational Linguistics" as a result of overlap with computer science, linguistics, or a discipline called decision support systems "Decision Support Systems"

The evolution of computer sciences was started since the discovery of computer laboratories and manufacturing in the number of American universities and the British in the early sixties.

The computer science was not as separate discipline as pure science or applied, but had its inception in the arms of scientific sections, especially sections of mathematics and electrical engineering departments.

As computer hardware and curriculum were not enough to be in different department at the beginning. The first computer hardware trend was related to engineering and architectural computer components. The second trend was the software programs, which examines a number of programming languages in science and engineering faculties used to solve the issues of mathematics computational and engineering and statistical purposes or modeling and simulation. The evolution of computers in terms of physical device "Hardware" and "Software" began to widen in scientific, theoretical and practical trends. This expansion has led to accelerated development of computer science as one of the basic sciences.

The large number of universities and colleges established the formation of separate departments for the computer science, such as, computer science, information computer, information science, computer education or cybernetics, or information processing, data processing. The resulting marriage of computer science with other fields of science, new scientific knowledge was created, especially in the field of application. For example, the evolution of computer "Computational Linguistics" as a result of overlapping computer science with the knowledge of language or field of knowledge called "Decision Support Systems" overlap between computer science or science information systems and management science and operations research and statistics.

The e-learning is intended to have several different types of education and training is different in terms of the nature of the educational process and content, methodology and fundamental difference between education and tele-education.

It is activated in this regard by the university of Baghdad programme of lectures by visiting Professors through e-education system which is being built in collaboration with the University of e-learning in Canada, where these techniques will open up prospects in the wider access to scientific developments at the international level.

The Quality Assurance and international partnership with Open University are applied by the Arab Open University. The most prominent features of this university are to lift restrictions on the admission and registration, contrary to what happens in most universities of open education in the world. The education system in the Arab Open University bounds student to spend time in mandatory meetings and seminars. This Open University believes whole importance of quality assurance and seeks first to measure the performance quality university and then make the means of quality support and assurance. It has adopted measurement tools and has evaluated the quality of British Open University and divided the university system to ensure quality and measurable ways and means of support.

The Arab and international experiences in the field of e-learning at the university began in Education in 2000. Characterized as an important educational issue in our contemporary terms by giving to science students the ability to search and investigation and find modern information. It also confirms that the scientists learning does not mean merely exploiting the potential of modern technology in the delivery and knowledge. Students can now attend the university located in places other than those that live in. accordingly E-learning includes concepts and philosophy, objectives, certain patterns, pillars, requirements and mechanisms.

In order to shed light on distance education and its importance, it is suggested the following points:
- The reality of distance education in terms of goals and ambition.
- The importance of the challenges facing the movement of distance education.
- Effectiveness of distance education compared to the traditional model of education.

It is also stresses that the Arab strategy for distance education in 2004 is to:
- Liberate human beings rights and maximize its contributions to progress of development.
- Meet the need for integration with other learning systems, including the education sector.

The distance education system is an educational model based on modern technology, but there are many obstacles facing the problem of application, such as, recognition, criteria, measurement, standards and quality and others.

The positive results on the importance of distance learning as model of education compared to other models, suggest thinking in different forms and regulations and renewable methods of distance education to suit the point in time of the third millennium.

Adoption systems of distance education through computer networks on the concept of the overall approach include a series of general educational measures in electronic form so as to be accessible to scholars. Distance education is one of the important means of communications and technological revolution in the transfer of knowledge and its uses to develop and employ them in the development of human capabilities and enabling environment for communication to the world of technology and information between individuals and among all sources of knowledge everywhere reach these networks and produces the network of learner direct contact with science continuously. There are also regularly available information, photos, and recordings through the web, along with holding meetings and symposia.

The Ministry of Education with the American Academy of Science to convene a special joint program, to develop university libraries and to qualify Iraqi digital age, to bridge the information gap experienced by universities and securing sources of information more sophisticated to either academic researchers and graduate students. This information includes teaching and working in the rehabilitation of university libraries and to familiarize them with modern information technologies used in the libraries of major scientific institutions of academic science is so-called virtual library.

The use of the Internet has become a reality impose itself as a modern method used in university education, but there are many things to identify clearly the benefit from its use and encourage its spread in future, including:
- Descriptions of output achieved by the kind of education.
- Clarification the methods of coordination between various education systems through the network.
- Development of assessment methods appropriate for students.

- Update instructions in line with the software used in line with the software used for this type of education.
- Cooperation colleges in the corresponding decisions similar to this type of education.
- License modernization so as to protect the rights of electronic publications.
- Determining academic standards for admission.

Techniques used in science

Most of the Bio-polymers with the electrical charge is transmitted in the electric field, which have the advantage to classify macroscopic particles and measuring the molecular weight and discrimination and diagnosis of amino acid changes with charges the components without charge, and vice versa.

The migration of the electric voltage low- lying useless to separate small molecules such as amino acids, either in the case of the migration of high voltage electricity will become more rapid separation of the amino acids, for example and using the electric gel migration and multi- acrylamide is the last for the time being one of the best prevailing circles to separate proteins and molecules with the addition of the compound. Dodecyl sulfate becomes possible to measure the molecular weights and nucleic acids and proteins, and at times used for this purpose the Agarose gel. The separation of DNA with a single when the use of acrylamide and Agarose. The use of other electrical relay with a multi- acrylamide gel to separate the circular DNA which access through the gel. There are electric displacement by the immune system, which can separate the materials that have the same movement with vertical orientation or by gromotography followed by deportation cuts.

Starch gel used for the first time in the migration electricity and usually consists of' starch pastes, potato starch which burned grain thermally and after putting them in the organizer and the creation of the gel horizontally, as noted in the shape of the sample in a small part which consists of pieces of gel using a razor blade and sealed with wax is usually the part or material lubricating and then begin operation and pass the voltage specified.

Migration of the electric type of the SDS gel, it is possible to calculate the molecular weights of most proteins, the measurement of movement in transition in the multi- acrylamide gel, which contains "(SDS) Sodium Dodecyl Sulfate". In the neutral pH and concentration of 1% of "0.1, SDS" of mercaptothenal. Most proteins associated with multiple strings SDS and analyzed sulphide bilateral ties by "Mercaptoethanol". The result is damaged to the bilateral structure of proteins.

The complexes consisting of units of high protein and the SDS and imposes the existence and status of helical and random act of proteins of this method, as if with a regular form, have an equal proportion of the charge / mass, owing to the percentage of the amount of the SDS associated with each unit by weight.

Using samples of unknown molecular weight of two known molecular weights can then calculate the molecular weight of the sample unknown degree of accuracy between 5% - 10%, and this is certainly one of the most well- known methods and used these days to estimate the molecular weight of subordinate units. It is noted that the way the SDS gel can be used to measure the probability of presence of aggregates SDS. When an electrical relay to a series of proteins of known molecular weights by gel column first separated into a series of packets and then draw distance movement of the samples against the logarithm of molecular weight together there is a straight line.

proteins meaning they contain both negatively charged groups and positively charged, depending on the pH is positively charged when the pH and negative charge in the event that the pH is high. In addition every pH in which the zero- charge and the so-called point equivalent to the movement. For a mixture of proteins, it may point equivalent to multiple sites called points equivalent, which consists of a pH gradient in a column containing the tube- negative and positive. A wide range of points equal to the charge and the various compositions is a mixture of polymers of carboxylic acids.

Spectral application

In order to simplify the regulations of life systems, spectral techniques are used to study the structure of many of the synthetic compounds with important skill of life, including protein, nucleic acids and others. Moreover spectral techniques are used in the follow- up of chemical reactions of life and employment regulations of life in the human body.

The techniques continued often its development in medical diagnosis, which can be linked to any disease, accurate diagnosis of the variables affecting the livelihood systems and chemical structure of various compounds, or monitor the status of one or more of the fabric that differ in normal and pathological cases. Molecules absorb light and length of waves that are absorbed and the efficiency of this absorption that are dependent on both the structure of the molecule and its surroundings, making it a useful tool for absorption spectroscopy characterization of small macroscopic particles.

Spectrum of ultraviolet and visible, the measurement of absorbance in the ultraviolet and visible for several purposes, including measurement of the unknown substance test of some chemical reactions and diagnosis of materials and determine the structural parameters of the macroscopic molecules and follow- up the transition the coil for DNA, example is the double helix as well as the titration of pH spectrum of proteins and to identify features some of the proteins by way of solvent perturbation and by difference spectroscopy and identifying revealed the linking of small molecules to proteins and the union of protein - protein and solvent disorder of nucleic acids.

Applications of infrared in the Biochemistry applications of modern infrared for diagnosis of mutual hydrogen in proteins and diagnosis of the number of hydrogen bonds and aggregates and measuring the common effective group and broken through the process of metamorphosis and diagnosis of tautorserism forms by and the union between small molecules such as riboflavin and protein. In addition the identification the carboxyl groups in proteins as well as determining the status of hydrogen bonds in proteins and multi peptides to measure the direction of groups.

Raman spectrum "Raman Spectroscopy": of the important applications of the Raman spectrum in the compounds life, study the mechanisms of differs automerism structure and different kinds of amino acids and discriminate adenosine mono phosphate and triple phosphate and their ionic forms in solution as well as the diagnosis of helix, beta related structure random coil of amino acids and determine the number of disulfide bonds in proteins and determine the number of double bases in DNA.

Fluorescence Spectroscopy: there are two types of materials used in the fluorescence analysis of macroscopic molecules; the first contained the same molecules and macroscopic materials fluorescent foreign insert. For example, there are three types of fluor of proteins self tryptophan, tyrosine, phenylalanine due to the fluoresce of the proteins resulted from these compounds, which could be used to study the changing perceptions of the enzyme by the positive correlation factor, assistant recipes enzyme active center and studies on the metamorphosis of the protein and the location of the tryptophars in the enzyme. In many cases the material could be added the fluor in the molecules composed as is being considered, either by chemical duplication or by simple correlation method. The name of the method which will reacted by adding particles when analyzed by outer fluorescence. There are several requirements for materials and when the use of external fluorescence:

– The material is strongly linked in a prime location.
– The fluorescence must be very sensitive.

–   Should not affect the macroscopic features of the molecule that studied.

It is possible that light energy absorb only when the molecule move from the lower energy to the top, such transitions in the diagram with vertical lines. When the molecule is initially irritating, it represent the excess energy that will be shaky as energy molecule in one of the levels of seismic and seismic energy appears as heat as a result of collision with solvent molecules (when the inflammatory molecule in solution and reduce the molecular, level to the lower vibration for S1).

In many cases, material could be added to the partial been studied either by duplication or by chemical simple correlation.

A beam of light with high intensity passes during the X- ray Beam Analyzer to choose the wavelength to the cause of irritation (eg, wavelength is absorbed by the material enough brilliant), after going through that light Irritated through cell containing the sample. To avoid detection the incident light is restored then the fact that emitted in all directions so that it can shine.

Spectra of circular birefringence "CD, ORD"

A set of techniques that serve to know the status of the molecule or macroscopic and interactions, including absorption spectrum provides useful information of this kind, but that the study of absorbency of polarized light in a spectrum of optical rotation dispersion ORD and spectra of circular birefringence "CD" The method for measuring adoption of the wavelength on the viability of active rotation of polarized light and absorb the difference excellence polarized light and the direction right hand and the left. The physical basis for each of the "ORD" and the "CD" are identical, and are in fact different to address the challenge of polarized light with optically active molecules. And cause to contain a very large part of the molecules of life on active duty visually Hence, the "ORD" and the "CD" has too many applications due to the fact that the spectrum of "ORD" and the "CD" of proteins and nucleic acids. The resulting from the spatial asymmetry of the components of the amino acids and nucleotides sequentially, but for the macroscopic particles it states each of the "ORD" and the "CD" in the structural studies of proteins and nucleic acids and proteins.

The pens and solids are used for this purpose at times, but the solutions are used in most of the time measurements for the "ORD" and the "CD".

Solution is placed in a container called a cell, and the device consists of the light source changes then the wavelength, and a system for polarization of the light, and system for measuring the polarization after the passage of light through the cell.

That the way to know the secondary structure of protein based on the measurement of the curves of CD, ORD experimental multi- peptide. As for the proteins they include measurements of the three main forms, alpha helix, beta form and the coil random. It was found that the spectrum of proteins and the type that gives the same due to the impact of side chains on the rotary power by force, as well as peptide that occurs at times because of the hydrogen between two amino acids and peptides that have multiple heterogeneous long chain of the same strength turnover in each component, such as those owned by small cascade. The side chains of phenylalanine and tyrosine and histidine and tryptophan to the spectrum of CD when they are in certain situations and the sulphide bridges give two CD bands.

Recent developments in the application of circular birefringence CD and optical rotation dispersion ORD advantage of these modern methods to know the status of the molecule or macroscopic and the interactions between them, as well as the structure of poly peptide and it is believed that the key of the structural knowledge of the secondary structure of protein based on the ORD curves and the CD. Furthermore, it is believed to use the test of changes on the status of the proteins by CD and study the changes in the structure of enzymes caused by the substrate, inhibiters that are related to enzymes, which can be illustrated by the CD spectrum for a variety of enzymes when the enzymes interact with the substrate and inhibitors of enzymes to help, as accompanying the metamorphosis of protein changes in the CD due to the loss of the structure of the alpha and beta and increase the spectrum of the components of the coil at random.

Nuclear Magnetic Resanace

NMR magnet is a spectral method that can provide sufficient information about the structure of bio-multicellular molecules and the interactions that occur between molecules, as well as molecular motion.

NMR spectrum depends on:

- Displacement of chemical "Chemical Shift".
- Fixed double "Coupling Constant".

Accordingly it used for the following applications:

- Diagnosis of chemical structures.
- Theoretical studies on the chemical tautomerism of displacement.
- Studies on the tautomerism composition.
- The dynamic characterization of chemicals.
- Characterization of the spatial structures of chemicals.
- Effect of solvents.

Technological developments of nuclear magnetic resonance technical developments have taken place in nuclear magnetic resonance study after the hydrogen nuclei (protons) and the nucleus of fluorine with high sensitivity and where the introduced technology transfer. Fourer the use of flashes of measuring the sensitivity of nuclei such as carbon -13 Bio- molecule such as proteins require nuclear magnetic resonance equipment with high frequency 300Hz and more, where it is difficult to analyze a device at low frequency (60Hz) or average (220Hz). New equipments of nuclear magnetic resonance frequency of 600Hz, these machines analyze many of the protein molecules and those of other techniques with the use of two-dimensional and three-dimensions.

New devices of NMR manufactured magnet after the entry of two-dimensional superconducting magnets, where the magnet save at -270 degree using liquid helium and liquid nitrogen to maintain the temperature of the helium and prevent it from evaporation. To distinguish between one dimensions and two notes that the spectrum of NMR magnet with a one-dimensional recorded displacement and pairing constant on the same axis while the two dimensional displacement is recorded the on two different directions. The three- dimensional recording the two perpendicular to each other and are getting a nice Dox-spectrum. The most important uses of the latest is in the field of proteins.

NMH magnet is a spectral method that can provide sufficient information about the structure of molecules and macroscopic bio interactions that occur between molecules, which can calculate the arrangement of atoms in the spectrum, and the hydrogen atoms (difficult in a segregation analysis of X-ray diffraction can be identified on site by this spectrum), It is spread in the macroscopic particles, can also test different atoms (phosphorus, carbon, nitrogen and hydrogen) separately. These can be applied in determining the spectrum of protein structure and enzymatic study of active centers and link small molecules to proteins.

In general, there is a difficulty in analyzing the macroscopic particles, the presence of large numbers of lines of spectrum analysis, which distinguishes itself lead to difficult to diagnose due to the large numbers of potential tackles of each atom, where the border of the spectrum characterized by complexity theory. As well as it would be possible to study chemical reactions and the effect of drugs on the disease and to study the changes that occurs to the water molecules in living cells.

There is great potential for the study of chemical reactions using a spectrum of protein and phosphorus in living cells has made a technical developments in this study. The nuclei of other important life such as sodium, potassium, calcium, iron, cobalt and nitrogen. Examples of phosphorus-use spectral follow-up of the heart and life processes within the heart and to stop the interactions of life organic phosphates components as evidence of changes related to energy. The specialists in the field of NMR spectrum nucleus of sodium, potassium and cesium due to the presence of sodium and potassium in the human body and play significant roles in the interactions that occur in the cell and pressure in the maintenance of fluid balance inside and outside the cells and the role of sodium in the transmission of nerve signals through the cells and the neurological contract and it can be completed as well as there are free and linked and contribute to solutions in the process of blood clotting and visual processes and muscle contraction, and here we began to record NMR spectrum of the nuclei of calcium and magnesium as well as nucleus of silicon, tin, nickel and other industrial significance.

A nuclear magnetic resonance other spectral method that can provide sufficient information about the structure of polymers and living on the interactions that occur between molecules as well as on molecular movement. The multiplicity of benefits and return to:

- The possibility of calculating the order of the atoms of the NMR spectrum because it theoretically can provide information for the purpose of this account.
- The hydrogen atoms (difficult in a resolution of analysis of X- rays), but it is possible to determine its location by magnetic resonance image.

Nuclear magnetic requires resonance spectroscopy of the radio waves of the circular disk of the magnetic field and the absorption of radio waves recorded. The sample in the tube between the B- polarization of the magnet A load (104- 105 Chaos) describes the spiral shape in a plane perpendicular to the electric field of the magnet that surrounds the sample. The sender broadcasts reluctant high and constant (roughly 108 or circuits per second).

And the others in the magnet or in the frequency of radio waves in part cause no change in magnet strength, either the change in the frequency of radio waves.

Nuclear magnetic resonance capability to provide much information about protein structure, where the constants of the spectrum sensitive to changes in both the arrangement and status, for example, in the absence of any fusion or federation is a complex spectrum of peptide or protein which is the sum of the spectrum of protein components, which are amino acids.

Spectrum indicates the enzyme in natural forms and (random coil), which can be calculated from the amino acids of the protein, it is clear that there are significant differences between the spectrum which is calculated so that, it is noticeable, for example, noted the occurrence of poles that occurs when the large chemical proton in amino acid. But in the end, or carboxylic acids, or when the neighboring nitrogen atom to the bond, also a small composed displacement, when the proton adjacent to the carbon atom of the peptide bond, or carbon or nitrogen atom of the amino acid closest 2, also generated a small displacement, when carbon or nitrogen atom in the amino acid closest, in addition to this, when the protein status of normal displacement, spoke of the distinct chemical protons some acids in the form of security.

Further studies in science

These issues raised by biotechnology as a result they have become at the forefront of basic research and applied consistently reached new levels of progress and complexity and can be far- reaching impact and positions that require scientific, political, moral and social. These issues vary with varying impact and could be referred to some of them:

- Cloning.
- Human Genome.
- Gene therapy.
- Map of the protein.
- Food and genetically modified organisms.
- Advanced technologies (nanotechnology).
- Vital information.
- Monoclonal antibodies.
- Biotechnology, medicine, agriculture.
- The discovery of disease- causing genes.
- Forensic – DNA, Fingerprinting.
- Biotechnology and biosafety.

- Biotechnology and environmental balance.
- Scientific strategies of biotechnology.
- Research and development and stops future.
- The role of education and training in biotechnology development.
- The twenty-first century is the century of Biotechnology vitality and prospects and challenges.
- Technical aspects of genetic engineering.
- Bioinformatics

It can focus on hot issues of the following:
- Cloning.
- Human genome.
- Gene therapy.
- Genetically modified food.
- Ethics.
- Genetic engineering and the internet.
- Biological weapons.

The third cloning occurs from mature cells, the researcher can wait to see the nature of the thing with his own eyes before they proceed on reproduction. This type of cloning is new and are using the nucleus of an adult cell adopted by Wilmot (embryologist) to clone a sheep from an adult cell, as well as it has managed to clone sheep, genetically modified with genes to produce human factors and blood clot of the most famous that carry human genes. Dolly, the most famous reproduced onganism that carry human genies while the cloned Dolly the first to reproduce the new technology. International principles for the reproduction of new life based on the cloning of a somatic cell is a mature steps each one of them have the status associated with these principles and is the following:

- The use of a cell donor (host) intended for reproduction of cells extracted from the membrane to view white pregnant sheep featuring the genes needed to form a complete sheep for the purpose of conversion to a cell capable of regeneration without preproduction and become a creature without full sexual intercourse (without vaccination) and instead of being cell embryonic viable fusion.

- Using an egg cell is extracted from sheep attached to another type of black- headed in the pot tester removed while retaining the nucleus with cytoplasm to put it right.

- The cell donors to famine and a halt to development and divisions, and preventing food resources for a period of ten days and give in to the state of sleep.

- The removed egg from the receptor cell near the nucleus of the cell donor prepared for cloning by using electric induction, small electrical firing bursts per nucleus and the egg unite and begin to act as fertilize.

- Allowing the egg genetic cells for growth and division and the formation embryo in the laboratory, implants in the uterus of sheep with black head and become the sheep that was born and was named an international replica of the white sheep of the cell donor and not the same as that used for the black bosom.

Genome

The human chromosomes is 46 consisting of strips of the double helical DNA wrapped circumvent complex shapes of helix, normal and high and consists of the DNA with four units of high repeated synthetic (nucleotides incomplete oxygen), each of which consists of three components: nitrogenous base and sugar phosphate penta and not organic.

The genome is intended to aggregate the DNA (genes) of the bacterial cell example, contain about 200-300 gene, while the genome of a human cell include the thousand times as much as the genes found in bacterial cell 200.000 to 300.000 and the organization of these genes depends on the number, in the chromosomes of the cell Eukaryotic (human cell) is more complex than the primitive cell nucleus (bacteria).

The genetic map represents the order of genes (genes) within the cell chromosomes and that this arrangement within the human chromosomes is more complicated than other organisms. Thus, the process of discovering how to arrange these genes, given the sheer number and complexity associated with variation built and responsibility to control complex cellular functions. Hence the decoding process and diagnosis and scheduling of full human genome by genetic map and the preliminary draft of a preliminary genetic blueprint of human genes and the previous process is equivalent to a significant scientific breakthrough, scientific achievements made during the twentieth century, including the discovery of penicillin and landing on the lunar surface and use a computer and other discoveries.

The discovery of the human genome and complete the approximate locations and sequencing of this large number of genes input to future developments are to:

- A new look for the human body.
- To find new ways of treating diseases (gene therapy developed) such as AIDS, cancer and heart disease.
- Correcting the genetic errors.
- Organ transplantation.
- To address the social and ethical consequences.
- Sustaining life.

Thousands of diseases due to the presence of genes responsible for the appearance and many of them dangerous to humans and non- treatment or cure and concerns and applications of biotechnology to find what is known as gene therapy, which is either by bringing the damaged gene or gene intact repair defective gene. This could be done through the intervention to repair the gene in somatic cells, or by intervening in the cell construction.

In order to spread the use of gene therapy it must be certain that the expiry date and free from damage and researchers that must be able to transport techniques and control gene expression in the correct and consistent, and should not obscure the international success of the many risks carried by this treatment.

- The genetic balance of any human being is the only thing that can not be replaced but must be preserved and transferred to the generations of while it is possible.
- Here we must make sure that it can allow gene therapy in somatic cells, and prevention must be in the manipulation because of its many negative consequences both in terms of genetic or moral.
- Gene therapy in somatic cell only affects the individual patient treat him, while affecting.
- Gene therapy in cells on the construction of successive generations.

It seems that gene therapy is the only way to cure genetic diseases or chronic (such as cancer, Acquired Immune Deficiency Syndrome). In this case, there is an objective one is to improve the health status or save the life of a patient work of the highly desirable, and then there is no difference between the unit body and the unity of the genes. Gene therapy holds great danger is the use of this technology in order to improve the human race.

It was discovered more than two thousand genetic disease; all affect the genetic information in the patient and move it to the next generation. In order to restore the natural functions of the victim there are two ways in gene therapy.
- Gene therapy in somatic cell.
- Gene therapy in the cell construction.

In each of these methods a special set of scientific and ethical considerations. Construction cells are sperm cells and egg cells, which include the rest of the cells in vivo somatic cells. The gene therapy of somatic cells in the introduction of DNA in this type of cells so the added gene in the progeny of the patient on the contrary, affects gene therapy in cells on the construction rights in the early stages of embryo development.

The structural gene therapy did not apply to humans, as it was rejected on moral grounds. But a team of researchers are currently considering the possibility of its application in the treatment of incurable genetic disease and it comes in all cases. The treatment depends on the addendum, any patient that the gene dysfunction and genetic causes of the disease will not heal or replaced, but added to the cell intact copy of it and this method of treatment do not apply except in cases of genetic diseases caused by genes elected. In general, gene expression is not valid only in a given tissue can be determined in different ways in experiments conducted on the living body, determines the quality of the input method the target tissue. Incomes through the trachea transfer of genes into the pulmonary epithelium, and injection in the liver gene transfer to liver tissue, either tumor injection in the objects is transferred genes into tumor cells also contributes to the carrier of the virus in determining the target tissue.

Different methods of gene transfer

- Viruses: Retro virus, including the only to ensure that the genes transmitted via cellular divisions. We therefore consider these viruses are the most successful means of gene transfer in the laboratory, where they allow in principle a final treatment of genetic diseases.
- Chemical methods: There are numerous studies on the possible use of fat bodies and compounds multi positively charged.
- Physical methods: The target cells is ejected shells are small technisten DNA speed due to electrical discharge or explosion compressed gas.
- Intramuscular injection: Move DNA the intentions of cells around the injection site but does not merge with it, but stay for extending from a few weeks to a few months to form ring.

New ways to design a treatment using cells of the organism
- Cell culture installed: is converted to the infected cells in organs or in the culture the appropriate gene, and then build a specific structure and in the infected tissues. Thus, cells expressing the gene and provide the onboard product withdrawal.

Diseases which are currently subjected to genetic manipulation

- Cancer: he application of gene therapy in cancer is not now aims to correct genetic defects in somatic cells, but sought to allow the introduction of genes to eliminate them.

- Neurological disease: This treatment is used to reduce nerve damage that accompanies Parkinson's and Alzheimer's disease and to enable the infected neurons to recover their function.

- Acquired Immune Deficiency Syndrome: There are many techniques under the anti- vaccination testing, such as installation and self-inoculation cultures for the primary fiber cells that contain transience's carry code of viral proteins and immunization procedure by genes which are to block viral reproduction.

- Gene therapy of developed AIDS, the first attempt of gene therapy in humans (1990), while the enzyme that remove amin of adenosine make the child is not capable of performing the immune response to resist infections after isolation of lymphocytes from patients, and then insert a normal gent for an enzyme that remove the amine through   virus vector.

Fingerprint

That science is progressing dramatically in the current year, so that it can be made in the last quarter of the last century, equivalent to human progress in its long history as a whole. In the field of genetics offers this impressive progress of science and builds on the many hopes in the future of human. While the human in a state of surprise and astonishment, which inherited the technology to adapt the results of gene.

Proteins play an important role in maintaining the structure of the genetic material that lead to reveal the full individual. The finger printing began until 1984, when the geneticist deploy at the University of Leicester in London search of genetic material that may be repeated several times and re-sequences itself incomprehensible represented in the length and location, this research reach in one year that this sequences characteristic of each individual and can not be similar between the two except in cases of identical twins only, and the potential similarity of fingerprints between one person and another one trillion, making it impossible similarity, that was found that these differences are unique to each person just like fingerprints and therefore called the fingerprint genes. Dr. Alec has recorded his discovery in 1985 and named them the name of the sequences of the human person as defined and as a means of identifying a person through the passages approach sometimes called DNA fingerprint "DNA Typing".

The genetic fingerprint known through the courts, although had spent time in the detection through forensic medicine, where possible knowledge of this fingerprint to identify the mutilated bodies and tracking children and missing soldiers, as it can mark genes to identify the person until the bulbs of the hair that has been cleared of many of the defendants by identifying the genetic fingerprint of murder, rape and revealed the true perpetrator of the crime, had a genetic fingerprint of the word on the issue of genealogy polarities of a number of issues to prove paternity, rape, and calculates the ratio of the distinction between individuals using fingerprint genes found that this ratio up to about 1: 300 million people, there is one person with the same genetic fingerprint was also found that fingerprint genes inherited according to Mendel's laws of genetics.

That this issue is distinct attention to enrich the scientific research related to this section of the scientific specialization, which is still growing and evolving as a number of questions that put precision together constitute the social issues and scientific issues that require a unique answer response send in self-certainty and a sense of security and assurance. The importance of this subject is first that it does not affect the religious, but very cautiously and in accordance with insights and analysis are limited, and the cure of genetic testing and abortion, infertility and human eugenics and other topics related to the needs of the Muslim scholars to discuss and study and comparison with the fundamentals of the faith and purposes of the law.

The gene therapy in somatic cells aims to treat serious diseases, and the possibility of morally acceptable. The gene therapy in cells construction remains a subject of controversy with regard to cell construction cells and with regard there are a number of questions.

- Do we have the right to change the genome of an unborn child?

- Who has the right to approve?

- Are we encouraging the introduction of genes (such as growth hormone) to improve the quality of embryos? Any non-therapeutic uses.

Chapter seven

Scientific Terminology

Preface

It is believed that Sumerians were the first who put the foundations in building a systematic record of scientific terminology. Then the Arab Islamic Renaissance used a lot of chemical terminology during their translation of science and philosophy from Greece. The names of initial elements of chemical used the Arabic language terminology, including escalation and distillation. By the age of industrial revolution in Europe entered the terms chemotherapy clearly formulated under the founded descriptions and characteristics of chemical substances.

In the mid seventeenth century the movement began in Europe's new chemicals (chemistry construction and installation), urea was synthesized from inorganic materials by Kohler the number of materials processed before 1828 even before the first World War, materials rose 200 thousand to 750 thousand at the end of 1983, then five million chemical at present according to AUPAC. The organization of AUPAC developed rules for designing a systematic manner. Reflecting on the chemical composition of Arab term, Arabs themselves have disagreed in Interchanger Arab. For example, if we take H_2SO_4, the German name it "Schwefel Saure" and the Syrian formula addendum sulphuric acid "Kibrit" and Iraqi sulphuric acid has been used by stokes system instead of accessories (ul, ur, ic, ou).

Assets of the scientific term

All of the world's languages participate in the three images to create scientific term, innovation and translation and Arabization, as in the Arabic language and ffatnama as in Vietnamese experience, but the Arabic language compliance and more driven to build scientific terminology.

Beside the Arabic language there are other languages, as well as further tests, where the Russian built after their revolution philosophical interpretation center of the establishment to transfer to the Russian language.

The experience of Vitamins approached in the college of medicine together with French experience as well as the Hebrew language which has been used in teaching of medicine, engineering. The Japanese experience in the transfer of science in spite of the alphabet amounts to ten thousand characters.

The scientific term is, for example with the following, as stated in the general information, definitions and provides:
- The difficulty of language structures in the face of scientific term.
- Abundance of media and Senate foreign language in the terminology in general.
- The capacity of diverse vocabulary to indicate when evaluating the term.
- The capacity of symbols and signals in the sciences.
- Diversity of diligence in the development of the term and the term science.

Oidal, which means (semi or similar) and then followed by Cardiac amyloidosis, amyloid, and alkaloid
- Not available for Arab students a comprehensive scientific dictionary is a specialized unit pronunciation and significance.
- There are a lot of precedents and foreign suffixes, roots and lack of clarity in the corresponding Arabic.
- The lack of buildings and Arabic versions that corresponds to the ratio formulas or affixing or description of the Greco-Latin and modern European.
- Other things such as not expanding the use of characters, symbols and signals Arabic.
- The lack of clearly written in brackets chemical formulas and the significance of these bows are generally set in the kindness of the arc to another group in another market or roots.

- Instability to write the names of most of the purely Arab Chemical elements and compounds except for some individual interpretations, including those by the late Sheikh Ahmed Alexandria, where he translated their names, an accurate translation into Arabic, including Almsdy of oxygen.

That although the symbols powered by decades dealt with Arabic Dictionaries Dictionary of honor such as for example when writing formulas for example, and not the general composition.

Recommended a number of Councils in the Arab language to use some of the rules in the development of new chemical terms, including:
- Attributed to the word combination when discrimination or to prevent confusion.
- When combining different source types, such as: radiation (collection of radiation).

The term chemical- issues and methodology compiled and passed those localized at different stages in dealing with the term chemical midst of it simple and compound term consisting of a few passages or words continuously, it also very complex because of the diversity of the names of chemical compounds, concepts and perspectives and sophisticated equipment. Accordingly, it has faced researchers in the development of the term different methodologies, including those dealing with the terminology in simple installation, which prefers to resort to foreign term systematic, which is characterized by semantic characteristics, of the harsh, kept its own identity, especially those which were made using the International Union of Pure and Applied Chemistry known short IUPAC, and those faced by traditional Arabic term which is characterized by generally devoid of the curriculum and standardization clear, constructive and accuracy of verbal meaning and flexible derivation.

The requirement is that the researcher includes the Arabic term translated into Arabic language, its sources that are less and devote language capabilities and the translation of Arabization of words and terms of chemical and not to neglect the complex terminology and nomenclature of organic compounds and inorganic, life science and pharmaceuticals. A systematic term, whether Arab or foreign characteristics which make them divided into groups, each of which holds a special status of case and accessories that distinguish them from others, and this is consistent with chemical compounds and phenomena, concepts and devices.

Several problems have arisen when building the chemical term, and has taken several measures to reduce the confusion caused by the use of wrong terminology, for example, when use of the word (previous – Glyc) in biochemistry Afro- Greeks Glyk, which means sweet in the original, expanded then use to subscribe to everything related to the term unilateral simple sugar. And derivatives of these former Glycemia; which consists of two syllables mia which means blood and glyc, dating back to the sweet taste.

This endeavor leads us to emphasize that the English term or other terms of foreign special what had been termed the concepts and the implications of scientific terminology may not be translated into Arabic as he may indicate, especially when some of the Arabic language that are not eligible for the expression of scientific terminology.

Construction and engineering the chemical term

Suggested approach for the formulation of the chemical term in the country of Iraq depends on the principle of measurement, and is characterized by certain characteristics lacked proper term, including:
- Easy comprehension and narrative to the original Bank.
- Lose to the Arabic phrase and subsequent elimination of controls.
- Regularly in the chemical groups common properties have distinctive marks of terminology.

An example is the Arabic term metabolic index, which represents the term transformative systematic international meta- bol- ism which consists of three sections.

314

Each chemical compound has special name (chemical formula) which is called chemical name subjected to rule and regulations to chemical, according to the system. Moreover, there is the name of the commons or normal "Common Name, Trivial name" may be the name commercially was launched the names of this information, and the names of almost supposed to be in accordance with the rules of Arabization names foreign language. This is not consistent with the views that have been localization of organic compounds, which represent the code parsed from the international code that has a meaning for example: "atri- ene-" is a triad where represents the three double bonds "edi- ene", which represents the parsed binary code where or where it represents the union of two double bonds.

The development of the following rules:

Where - ane- en if you enter on the name it means that the compound is unsaturated (ie. the double bond) C = c

Propene $CH_3 - CH = CH_2$

Butane $CH_3 - CH_2 - CH = CH_2$

Cyclohexene – ine

Is that the compound with the three- bond, C= C

Example:

OL

If you enter this to the name, it means that the compound alcohol, which contains a group (hydroxyl- OH) for example:

Methanol $CH_3 - OH$ Methanol

Ethanol $CH_3 - CH_2OH$ Ethanol

Cyclohexano

If you enter this to the name, so the compound becomes aldehyde, which contains the group – CHO group is called aldehyde, and read his name with the subsequent example:

CH_3. CHO Methanal

CH_3CH_2CHO Ethanal

5- one

This is called Ketonic group

If the C = O on the chemical name is said to read with the subsequent

Example:

$CH_3.CO.CH_3$ Acetone

Aceto phenone

6 - ose

If entered to the name composite sugar for example:

Pentose sugar

Hexose sugar

Mannose 7- (- azo)

If entered this group (-N = N-) in the name we got on the chemical compound called azo

8- ide

Used to express the derivatives, was named the compound adrenaline

1-[3, 4- di hydroxyl phenyl 1] = 2 – methyl 1

aminoethanol

Accordance with formula purely foreign words, any word Arab is the following:

1- [3, 4- Dual- hydroxyl Phentl] – 2 amino ethanol instance

2- Methy – 3 – phenyl- 1, 4 naphyhaquinone

Majeed Qasis has proposed special conditions, including:

• Follow-up to international norms in the Arabic labels.

• Retention of cases and accessories and international symbols of the wording and meaning.

• Maintaining a vernacular names and names of Arab heritage.

• Transfer of the vehicle under the foreign name of buildings and the use of Arabic language symbolic images.

Translation

Translation in the Arab-Islamic times

Methodology adopted in these covenants on the developments that occurred in other countries. The translation contributed to the promoting

Arab-Islamic entity covenants, which began in the first century of migration and ended in mid-fourth century.

The translation passed stages started from extrusion specifically in the Umayyad era to its end around the middle of the fourth century. It has appeared to us that the Abbasid period alone was characterized by the emergence of more systematic methodology in the translation movement, including the systematic phase of Mansur and methodology. The next stage was the era of Al-Mamoon that lasted until the middle of the Fourth century of the Islamic Calendar (started with the Prophet Mohammed migration).

Therefore, the development of methodology for the translation movement may represent the four main trends and Al-Mamoon in the Umayyad era in the age of Mansour, Rashid and translation in the era of Mamoon. The methodology of the especially the first Abbasid era depends on:

- Fluency in Arabic.
- Conservation of the stock, verbal foreign language.
- Definition of foreign terms on the structure of the Arab tongue.

Translation in most cases is considered using translated art forms individually and others tended to convert the translation into science. What supports the process of translating the language itself is characterized by key features of grammar and morphological settings and the fact that translation is central to deal with two languages that have the advantages of each. Translation accompanied by the growth of human groups and collected them throughout different history links, and about the differences of these groups in the language that had to be a bilingual to secure the understanding among them. Assyrian King Sargon has published in many languages, as were the people of Babylon during the reign of Hammurabi, 2100 BC. using multiple languages. Rashid Stone gave the key to the mysteries of Egypt and save the text of a treaty between the Egyptians and the Hittites in the Egyptian and Hittite language.

Schools for the training used over time, the translators and many of the researchers in many languages ideas regarding the process of translation. But these ideas did not come out of the theory in translation studies, or clear the issues raised by this theory. Therefore, the difficulties facing the formation of the clear theory in the translation mainly by the multiplicity of kinds, including for example, word for word translation and literal translation, faithful translation and translation of the moral, free translation, translation of conventional interpretation, translation service, prose translation of information and knowledge and thus academy translation. Then it is difficult to develop the theory and a clear methodology and parameters to deal with all types of translations.

The Assyrian King Sargon of used the many spoken languages, the people of Babylon during the reign of Hammurabi, about the year 2100 BC. Multiple languages and translators employed by the ancients to the creation of dictionaries, preserved on cuneiform clay patches.

At approximately 2400 BC., Gaius Andronicus translated Homer's Odyssey, a poem to the Latin language, and also quoted Tivuos Aionnbos a number of plays to the Greco- Latin languages. Thus, the methodology focused on the terminology of poetry. The Rashid stone the second century DC, has been known to write the key to Egyptian Hieroglyphics and wrote two forms based on a systematic basis Aldematip language and Greco-Egyptian. Jerome follow in 284 AD, the principles of translation that involves the transfer of meaning and sense, but not systematic transfer of word and by this stage, it was so- called translation with no methodology.

This research has formulated the systematic translation through the followings:
- The use of terms to help pottery connectivity.
- Use vocabulary and expressions neglecting the specialized vocabulary.
- Use special phrases when necessary.

Played a clear role in the translation during sixteenth century and he establishes the theory of special courses and special rules included in the recipes below:
- Knowledge of the language that translates it and, which translates to it.
- Knowledge of the contents of the writer.

- A way from the translation word for word.
- Use formulas to speak in circulation.

Dryden said at the name time not to carry out the translation of characters where they are as described by dancing on the ropes does not lead to clarity of vision.

It has been shown in 1790 by Taitler methodology for translation from the other principles, including:
- Writing style with similar to the original.
- Translation mode through strain of originally.

The methodology for the translation as art has dealt with many researchers, including Dalembert and Aboboto and Dr. Campbell Hulusi that believes Dalembert's remarks can not be considered rules or even principles of the art of translation, while the Abu Bhutto developed basis for the methodology of grammar, Campbell also presented the criteria for good translation methodology included the following:
- The clear language of the original text.
- Transfer the spirit of copyright and its ways to the text of the interpreter.

The benefits that can be obtained from the study of translation as a methodology art is represented:
- Comparism of the language with another language.
- Strengthening the capacity of writing in two languages (the sender n11receiver).
- Auditing the terminology and the development of rules.

And can be summarized in the subsequent stages that we have mentioned:
- Jerome and principles of translation.
- Luther and concepts of the new translation.
- Dolly which developed the first theory for translation.
- Titlei with his extra counterpart.
- Dalembert and Aboboto and Campbell Translation as an art.

Literal translation

Characterized by translating the original text word for word, which made it necessary to put the word translated as foreign, causing in the receipt of many of the terms of non-Arab, which remained in use until the present.

Translation sense

The translation was carried of the sentences together but not separate word. Han Ben Yitzhak Abadi was considered as the most famous who followed this method of translation.

Assessment of the two methods

The first method as a suitable as a language school for students word without meaning, either way, but the second way draws the meaning and to transfer the language-borne sound. The translations which were made by the Siryac Greek science, which were transferred accurately and honestly such as, medicine mathematics with the occurrence of the circumstances from the Greek into Syriac and Syriac into Arabic. Despite the fact that translators Hanin Ben Yitzhak Abadi and his son, Yitzhak Ben- Hanin and a nephew and Habeechebn bin Thabit bin Kasna and Luqa, and Jacob Ibn Ishaq Kindi, concentrate on those of the famous capacity of knowledge and understanding of the subject.

The prosperity of translation was due to the care of the caliphs to translation for some families to the translation and the desire of some ministers and doctors access to the ancient civilizations.

The features of the methodology of this era with the following:
- The need to key factors that contributed in the translation into Arabic.
- A literal translation which was led by John the son of Al- Batrig.
- Sense of translation, mainly contributed by Hanin Ibn lshaq.

- The existence of stages in the translation such as an implicit way of translation from Greek into Arabic language through mode of mediation is Syriac.

There are a number of technical difficulties of translation, including:

- Translation of the examples those are likely to more than one meaning.

- Translation of the new words or new developments and metaphors and words that are not in the lexicon.

- Lack of technical terms, the interview in Arabic.

- Risk of some linguistic terms.

- A choice between foreign and used the word in Arabic and Arabized word.

- Duplication of language and sometimes triples, quads and a good example of the translation of the Koran where the translator is a serious mistake in the translation of poetry or losing beauty.

According to the standard of NIDA, in describing the process of translation and understanding of the source language (language sense) it is necessary for the translator to be able to unscramble voice or written, and to understand the meanings of words and expressions in the context of various reforms. The requirements of expression in the target language is represented by the translator by using voice tags and use of knowledge and words and conventional expressions used truly and the use of grammatical structures used morphological true.

Translation assistants and their means

There are two types of aids (printed reference) the first is relating to the understanding and other by expression. Examples of such assistance are: monolingual dictionaries (verbal dictionaries), spelling, grammar and cultural references, as well as geographic, historical and social knowledge references to help the reader, or the interpreter as well. The Auditor Professional is also used in the translation of scientific, cultural and doctrinal, including dictionary, definitions and keys to Jarjani Sciences and the Al-Khwarizmi may be monolingual dictionaries or bilingual dictionary, such as technical or scientific terms.

The expression assistants are used in the field of expression in the target language or language standpoint include language dictionaries and references that include bilingual dictionaries mismatch of public and private and the last under the terms of economic, sociology, medicine and engineering.

The methods of translation are intended to machine tools (EDP) equipment, which invests them dictionaries; translation machine automatic banks terms database data in the translation and e-mail. In this area it is believed that Frepts Bepkp out of focus on the value of a human translator, it is believed that the work of electronic machine translation can not be called a translation of it is the task of the transfer of meaning without understanding and groping in the text of the relations of language inside. Nayda, believes that the aid in the translation is not that simple, but transfers itself directly on the surface of the text, but the level of analysis into three phases that include the full texts related the target point and target language texts related closely in full.

- Coordination of the texts.
- E-mail.

Frepts Bepkp believed also that the work of electronic translation machine can not be called a translation but it is not based on understanding where that:

- Machine is the task of transfer of meaning only.
- The translator of human has the capacity to understand what the relationships in the text.

Nayda confirm as we stated the translation processes, is not transfer directly on the surface level of the text, but it is analytical processes based on synthetic human capacity in the first place.

Scientific Translation

The scientific text characterized by special features including:

- Descriptors feature: It provides features by writer and not self-descriptive of the scientific truth. The author writes of scientific text directly depend on the coherence of its impressions of self.

- Formal feature: The language of the scientific text and way to deliver content but it is not an end in itself and the form of scientific work is not an integral part of the content.

- The Oneness of meaning in the scientific text. Scientific meaning of the text is likely of one explanation.

- Lack of scientific text of linking with defined time. Skip business major scientific barrier of time and place.

The distinctive features may be invested of the scientific text, reliable and scientific texts translator seeks to communicate a message with aesthetic though the aesthetic quality the desire for greater clarity and precision of expression.

The translator of the scientific text adapt what he meant with the nature of the message that it is delivered, the text of scientific education in the first place and where some of the aesthetic values.

A translator of scientific text needs to acquire daily a huge amount of new terminology, while the writer is moving in the area much narrower than the field of science, whatever the language, whatever the imagination because its development is much less rapid. The scientific text in some cases carrying an aesthetic adds to the content. The translation of the text distinguishes between scientific and literary translator and determines the position of each.

The translator of scientific text must be objective and should be committed to accuracy and to convey the text as evidenced as much as possible from the secretariat, taking into account the order of the elements of the sentence the same way as arranged in the original text. It is worth mentioning that to remember that the effects that may result from the wrong translation by way of using a drug or operation of an electric device. On the contrary the literary translation characterized by scientific innovation and creative means to a large extent the ability to imagine.

The importance of scientific translation increase as a result of the explosion of knowledge, and vast technological progress in all areas of life and is worth mentioning new to the Arab Human Development Report for 2000 that the total has been translated since the establishment of Dar al-Hamah in the era of Mamoon so far, an estimated ten thousand books.

The scientific translation became the most important demands of the scientific creation of Arab societies to engage in a knowledge society of the Arab Organization for Education, Culture and Science is seeking to develop an Arab strategy for translation and publication of major books of scientific culture, including scientific foreign sources involves both the translator and publisher.

This translation requires a set of infrastructure including dictionaries, terminology banks, and qualified institutions for the translation, the translation of scientific institutions requires of the media system centered on the role of media in spreading scientific culture, as well as information technology, which rely on machine translation between different languages as well as the indexing mechanism, and the summary and automatic building of electronic dictionaries. It is problematic translation of scientific aspects of copyright and intellectual property of the authors of the original texts.

The theory and scientific of translation and concepts

The translation is process with a language as a top transfer (message) between the two Parties, the sender, recipient and witness of the twentieth century in particular, the tendency to theorize the process of translation, on the basis of rules had to be a researcher of the theory of translation and thus prevent the translation into the science of translation is aware of recent origin, it is natural that different views of the workers.

Aravin Kochmidr says in the details of the scientific translation:

• It is not a transfer of substitution and the transfer of language (one to ones).

• The translator to discuss the components of the language accordingly.

Kochmidr illustrate planning translation process steps of the two consecutive operations in time:

• Diagnosis of the text in the language of spirit.

- Re-configures the text in the language of the target.

Kochmide confirms that the translation process occasionally requires a few modifications in the statement focused on the composition of the sentence or semantic features.

Translation approaches

These approaches are divided into:
- Technical curriculum, they are basically the following stages:
- Analysis of the language of the source and the receiver.
- Examination of the text in source.
- Regulatory approaches is t represented by using the regulatory framework for translation as follows:
- Liberalization of the first draft.
- Review the first draft.
- Read the translation to the achievement of the creative style.
- Studying the reactions of the recipients.

Nobirt divided generally the texts to be translated to:
- Texts related to language of point of view.
- The language of texts related vein target and roll together.
- Texts closely fully inked to the target language.

The division of Boser on a tripartite basis:
- Texts that distinguishes content.
- Texts which distinguish the shape.
- Texts which distinguishes impact.

Science and the theory of translation

Phyllis Volgeram believed that the task of science is to describe and explain the focus of a domain in which this science in a media analysis by using methods appropriate to the purpose. It consists on the translation of the three areas:

- Analysis of the source of the language accordingly.
- Transport between the two languages.
- The translation process and their product.

The task of the science of translation is an innovation approaches to these three areas and determine the conditions of conformity on a scientific investigation of safety.

Eugenio Cocherio finds that the difficulties encountered by researchers in establishing the theory, of the science of translation due to ambiguity of the relationship between the theory of translation and text linguistics, where the transport process of reaching on the basis of conversion from one language to another and the translation process of converting a linguistic and translation theory in its current form based on the concept of the just.

Antoine Popovic believes that the theory of translation is part of the following concepts:

- The general theory of translation.
- The special theory of translation.
- The practice and teaching of translation.

Literary translation organized by the specific theory of translation, which takes into account the various scientific fields

The translation may reflect the face of the writing linguistics as perceived by many intellectuals, where it was:

- Knowledge or can be art.
- Does not express the thought of the needs of the owner.
- The translator to know the languages of the transferee and the transferee to it.
- The reader considers the text after its stability in the new language, and then decides on the translator.

As stated in the above, are considered the question of the identity of the text in the languages, constitute the main difficulty in the way of the theory of translation.

The Mounin thinks that the translation is friction between languages, but an extreme case of friction between the cases and suggests that the teaching of contemporary linguistics and should be concerned with questions of translation for the following reasons:

- It is not permissible for to continue to ignore the translation movement.

- The translation on the behalf of contemporary linguistics.

- Mounin confirms that the art of translation is not limited within the boundaries of linguistics, but on the different faces.

- Mounin displays at the same time theories of linguistics that may deny the legitimacy of translation.

- Mounin concludes opinion on the concept of translation (all we can be connected from one person to another in a language that can be connected from one person to another, from one language to language).

There are other criteria that are referred to by many researchers in the translation of standard Marur, which states:

- Transfer of meaning in the translation.
- Transfer of structural appearance.
- Transfer stylistic appearance.

This criterion has indicated two questions the terms and composition should reflect and be a definition of the word is something that Bulslev. The terminology raises the question of taste or irrelevant.

Arabization of science

The issue of science Arabization is one of the most important issues facing the people of the Arabic language in this era characterized by accelerated scientific and technical progress, using different languages but not including Arabic. Exhibit the progress of invasion of the Arab language in some features. Note that the English language as well as the French

327

became the languages adopted in many fields, including the areas of science and technology.

Public education and higher education in Iraq are arabized in most fields such as law, education, economics, sociology, political science, science except medicine. The teaching was in science colleges in Iraq, in lecturing discussing, testing the English language was professor's study being tested. The Arabization of science had many grounds, including:

- The Native language is the title loyalty to the homeland.
- There is a feasible terminology enough for the needs of Arabization.
- Arabic language is manageable for example:
- Many of the weights derivatives (the names of machine).
- Standard pronunciation of one in particular time and place.
- Provides weights for the interaction process (attraction and interaction).
- Industrial source terminology (sensitivity).
- Many of the voiced are not available in English.

There are some who oppose the Arabization of pure science, which based on the perceptions of this particular language, including:

- Disqualification of the Arabic language in teaching science.
- The language of science and technology in the present era is English.
- The number of books and references in Arabic are a few.
- The teachers who are able to teach in Arabic are a few.
- Wondering of some of the terms issued by some Arab language academies.
- The problem of scientific symbols and letters and numbers.
- The problem of publishing and distribution.
- The mastery of a foreign language is essential for those engaged in science.
- The use of English in higher education raise the student's ability is most widely spoken language.
- The teaching of any foreign living language is aimed at mastering basic skills.

The permanent office of the Arabization carried out the referendum in 1966 on the validity of the Arabic language for university education. The following problems and obstacles that have emerged then and which continue to raise concerns:

- Failure of the Arab scientific research.
- The difficulty of the Arabic language in terms of the rules and writing.
- Lack of adequate references in the Arab areas of science.
- The scientific innovation and writing in Arabic are not encouraged.

There is no country in the world that did not adopt the national language in education. For example:

- A Chinese experience: was able to unite the Chinese language as it was divided to the 300 language, contains 44444 sections then canceled the dictionary of the English language in various kinds of science and technology.
- The French experience: France got rid of the teaching science in English; then has taken its Francing science until the French language has become a universal language to learn medicine.

Arabization means creating opposite to Arab foreign words and adaptation of foreign word somewhat similar to the Arabic words with alphabet sounds. The Arabization of linguistical means according to a Lisan Al- Arab by Ibn Manthoor the "Arabization for the Arabic Word".

The Arabization is currently intended to use Arabic education for all levels and scientific research in its various branches, specialties. In additions Arabization actually is intended by the dominance rule of the Arabic language on its territory and the language of science and education. The old Arabization intended to keep the format of foreign word to Arab tone, note that the Arabization is language of source the act Arabs and Arabs means expounded in the sense and Arabs, if a man spoke cast his argument.

The meaning of Arabization is intended to utter a foreign word language on a platform of Arab speech and weight (antidote and oxide). The Arabic language academies of assert the term, when agreed Arab word with the votes and weights. As well as the Arabization of the txt where the text is transferred from one foreign language into Arabic and Altjeem transfers the next form Arabic into foreign language and also make the Arabization of the domain where the language, such as Arabization in education. The translation is defined according to the Lisan Al-Arabs (translated speech is transmitted from one language to another language) and dictionary Al- Waseet (translated words between him and been underscored) and should differentiate between the diverse types of translation:

- Interpretation.
- Translation.

There are many examples of Arabization, including:
- Mashreqi form: has been characterized by this form:
-     Verbal attribute.
- As specialized technical precision.
- Moroccan model: it means the process of Arabization comprehensiveness and universality of all activities within the community.

The elements of the Arabization process consist of:
- Qualified prof. to teach Arabic.
- Scientific expression and the term.
- Book.
- Foreign language.
- Lecture.

Arab efforts to Arabization

There are varied experiences of Arabization in the Arab countries depending on individual circumstances of each country; including in Iraq is a university education in the Arabization of laws and decisions that led to the Arabization into force. Proponents of Arabization, the national language teaching aims to achieve the following:
- Combination of originality and contemporary.

- Strengthening the binds of the link between educational institutions and society.
- Standardization of culture and scientific and intellectual effort.
- Helping learners to understand and absorption.

The result of the delay in Arabization is the use of distorted English language in most cases and a lack of understanding of what is given to the students such as the information, embarrassment and the impasse in which it resides when the student is graduated from the non- English university.

Arabization conferences and symposiums

The most important decisions made at conferences or responsible official bodies for the following:

- Resolutions of the Conference of Ministers of Education and Higher Education Arab/ Morocco 1970 and demands that all Arab countries take the initiative as soon as possible to take measures and means to use the Arabic language teaching in all phases of general, vocational education and university levels.

- A decision by the Third Conference of Arab Health Ministers/ Cairo 1974 No. (4), which includes the use of Arabic in medical education.

- Resolution of the Fourth Conference of Ministers Responsible for Culture in the Arab World / Algeria 1983/ held under the slogan about the Arab cultural security and contained the decision to ensure that the classical Arabic language should be used for education at all levels and to be the safety of the language.

- Resolution of the Second Conference of Ministers Responsible for higher Education and Scientific Research in the Arab/ Tunisia 1983 / and includes:

  - To speed up the national and regional Arabization.
  - The integrated policy of Arabization of the Arab countries.

- Resolution of the Supreme Council of the GCC countries of the Gulf at its sixth session/ Muscat 1985 / and provides (the obligation to Arabization of higher education and university education in all its branches and specialties whenever possible). It includes:

- Adopting the recommendation of the meeting of heads and managers of universities and institutions of higher education in the Arab Gulf States about the commitment of Arabization, and a timetable and detailed plan for the Arabization of higher education.

- Formation of a special working group for the Arabization of higher education consists of representative of universities in the Gulf Cooperation Council and representatives of other relevant parties.

• Council of Ministers of Arab Health/ Khartoum 1987 numbered 10/ these include an operational plan and realistic for the Arabization of medical education in the Arab world.

• Resolution of the Council of Arab Ministers of Health- the thirteenth session/ Oman 1988 numbered 13 / these include:

- Raising the medical Education in Arabic to their respective governments for approval.

• Resolution of the Executive Office of the council of Arab Health Ministers of the fifty 1988 Geneva, for study of the experience of the Syrian Arab Republic in the field of Arabization of health education.

• Resolution of the Executive Office of the council of the Arab Ministers of Health in the session of Tripoli 1988 / completion of course a policy of Arabization health education.

• General Conference resolution XII of education ministers of Arab Gulf states, Kuwait 1993 and urged the universities to provide the resources and administrative structures necessary for the implementation of Arabization and supervision.

• Decision of the Conference of Ministers of education in the GCC/ Riyadh 1986.

Chapter eight

Science Horizons

Development of Science policies

The reality of science and technology in some third world countries is clear as follows:

- There are no clear scientific and technological policies.
- The formulation of policies does not include the subject of extensive discussion, including producers of science and technology and their beneficiaries.
- Lacking the participation of the private and public sector in formulating the elements of science and technology policies.
- Follow up research and development activities is nor serious.
- There is no discrimination of priorities such as food security and health services.
- \the specific modalities of implementation are limited especially from policy to the strategic operational planning.
- The advantage of international and regional initiatives is limited.
- Efforts in scheduling strategies of science and technology are sporadic and disjointed.

The policy of scientific and technological, the developments in the United States is adopted on the following:

- Addressing global challenges and the new national directly.
- The government imposes a major role in assisting private companies to grow.
- Economic growth is the focus of policies:
- The ability to competitiveness in the industry.
- Job creation.

The science and technology policies as being the development of a range of sectors covered by some direct scientific and technological activities and their relationship to the establishment of infrastructure, national science and technology. The wide range of investment financial policies relating to intellectual property rights, trade policies, export, transfer of technology and research and development.

Policies adopted by governments in departments and institutions in general contribute to national innovation, whether based on the technological capacity to local or imported technologies. However, these policies may be unrelated, but can sometimes be contradictory and therefore the main task of science and technology policy provides a framework for policy coordination and greater collective impact.

After reviewing a number of science policy and technology in the world is clear found to operate by the institutions, centers and independently on the assumption that many developing that many developing countries, including Iraq that does not have the actual potential for the use of science and technology in the development process. This will help in assessing the effectiveness of scientific institutions and technology as well as create and activate national capacity to coordinate the contributions of those policies in the development and adopt the criteria as follows when evaluating science policy:

• The presence or absence of science policy.
• A science policy is not effective (the inability to use science in development).
• Assessing the effectiveness of scientific institutions and centers in general.
• Creating an environment technological innovation.
• Coordination of technology between government departments.
• Working in partnership between industry and the federal government and universities.
• Focusing on new technology (information, communication, manufacturing technologies).
• Consideration of basic science is the base upon which all technological progress.
• Directing science policy toward the user.

The main following targets of the represent general framework for science and technology policies in Europe:

- Acting as a collective actor in the field of science and technology:
- Encouraging the use of research facilities better.
- Confirm the international role of European research.
- Strengthening the scientific capacity and technology in European countries.
- Development of the knowledge base.
- Enhancing the competitiveness of European industry on the international level.
- Narrowing the gap between technical areas and disadvantages areas in Europe.
- Meeting the social needs and economical for European Union.

Educational reforms are intended to subject education in radically revising based on the comprehensive analysis of the educational system according to variables in multiple things, including:

- Growth of knowledge on learning and the nature of learning.
- Rapid change in social and economic life.

The programs adopted in this issue of education reform for the knowledge economy, which has several components, including:

- Contributing to the government and administrative reform in educational policy guide.
- Renewal of learning programs to achieve the learning outcomes appropriate to the knowledge economy.
- Providing new environments for learning through the establishment of new school buildings.
- Contributing to the modalities for early childhood education to promote readiness to learn.

The education reform is contributed mainly on educational planning and the relationship between the two diverse, sprawling, multi- faceted, often branched issues, as the reform of a complex and difficult, therefore according to this relationship it requires a broad understanding in the aspects of planning as the main forces for reform and educational reform need to ask planning.

Moreover, the work on the reform of human development requires the education reform based on investment facts and information available.

In 1983 this report was issued an important document for education in the United States, which insisted that the problems of the U.S. nation in education is due primarily to lower academic standards for students and low quality of education and also pointed the finger for the teacher, after a 1990 report another, entitled America in 2000 education strategy included a lot of trends in the reform advocated by the report of 1983 and the present report contains the following:

- American student to occupy first place among the world in mathematics and science.
- The U.S. acquires all the colors of the knowledge and skills necessary to compete scientific and exercise of the rights and duties education is not a benefit to live, but for life.

Education strategy in the United States of America (2002 – 2007), Goals and objectives

- Equal access to education for each nation (do not leave any child out of school).
- Improvement the achievement of the student.
- Development safe schools.
- Converting to the field of education based on the facts.
- Raising the quality of adult education.
- Building a distinct administration.

- Linking federal education funding with the results.
- Increased flexibility and local control.
- Increase information and options for either parent.
- Encouragement the use of scientific methods within the federal education programs.
- Emphasizing that all students from third grade who are able good reading.
- Improvement the achievement of math and science education to all students.
- Improvement the achievement of all high school students.
- Improvement the quality of the teaching staff and school administrators.
- Ensuring schools safe and free from drugs, alcohol and cigarettes.
- Building powerful features and continuing to renew the nation's youth.
- Raising the quality of funded research.
- Strengthening the accountability of the post-secondary institutions.
- Reducing the gap in access to and complete college education among the masses of different students in terms of race and economic status.
- Building effective financing mechanisms for post-secondary education.
- Strengthen black colleges and Hispanic and indigenous peoples.
- Financial integrity and management and internal control.
- Improvement the strategic management of human capital.
- Resource management the technology and the use of e-government to improve the services of the ministry for its customers and parents of students.
- Modernization of the financial aid programs for students and the reduction of the fiscal deficit.
- Access to the integration of budget and achievement.
- Raising the participation of society and community-based organizations to increase the effectiveness of the programs of the ministry.
- To strive to become the ministry of high efficiency and organization based.

Evaluation of scientific research in Iraq

A sound scientific planning for the development of scientific research of the university requires first the study of the current reality as it is a scientific study designed to learn the possibilities available in the universities and was done in this area to identify the negative aspects of it.

The scientific research at universities in Iraq, is one of three core functions of teachers as well as teaching and community service is linked with the promotion of academic holders of doctorates, while the remaining percentage of teachers holding a master degree of the operation full-time faculty.

There is a constant increase in the number of teachers in Iraqi universities that commensurate with the continued increase in the numbers of students enrolled in universities, and the indicator on the high ratio of students to teacher, an additional burden on teaching, which in turn reflected negatively on scientific research quantity and quality in universities.

The postgraduate students in Iraqi universities are an important part of the manpower involved in scientific research. Those students shape the future researchers, at universities in developed countries and have an important role in the completion of scientific research. Despite the expansion of universities in postgraduate programmes in Iraq, their contribution to scientific research is still pursuing their studies and specification in difficult circumstances. The current reality of manpower in the relevant scientific research in Iraq is characterized by low numbers of full-time researchers in spite of individualism to conduct research and the scarcity of complementary research teams.

The high proportion of the number of students compared to the teachers of acknowledged international figures, and the scarcity of opportunities for research assistants and technicians for training in developed countries to deal with the specialized instruments and maintenance of research laboratory, is a problem that should be considered seriously. The available statistics indicate that Iraq had spent a few sums of money on research and development after (2003). In an analysis of the budgets of universities refer to focus on supporting research projects, publishing and missions.

It is well known that the funding of scientific research in a university in developed countries comes mostly from the industrial sector and has reached the proportion of the funding year (1996) in countries like Japan to 67% and the United States to 63%, while the contribution of the industrial sector in supporting scientific research in Arab countries is very modest, almost Non - existent in many universities. Statistics indicate that the total expenditure on research and development at universities and colleges in America (2002) to 36,333 billion dollars when the federal government contributed.

The productivity of doctorate-holding teachers in Iraqi universities in (2003) was 0.44 researches for each holder of a doctorate each year, while it was 1.00 a year ago (2003). Ministry of Higher Education in (2002) in collaboration with universities strides for the issuance of specialized scientific journals in nature and in various scientific disciplines and it is hoped that these magazines with a scientific level are classified and adopted globally.

The nature of research and investment that may lead to creative results after the year (2003) and industrial applications and patents are rare. It is clear, that contribution in this area is very modest, because of the inability of the institutions of higher education to transform the results of scientific research to investment project because of several obstacles. Iraqi universities after the year (2003) have received a package of legislations that have given special attention to scientific research. Instruction were issued own encouragement, regulations adopted by all universities and research achievements as a condition for promotion of distinguished professor status. The legislations included those issued by the university foundations to support scientific research, publishing support and instruction to published journals by the addition to patents and instruction: for the dispatch of teaching attending conferences and symposia.

Iraqi universities and scientific research

Scientific research in universities is inseparable about the problems of Iraqi society, where research aims to obtain a degree or promotion, its role in the production of knowledge and solving problems is very limited, and worth noting that universities operate on the developed world.

- Scientific research and its role in the production of knowledge and solving problems.
- Development a clear policy for scientific research.
- Availability of information to help scientific research.
- Service sectors of society productivity through scientific research.

The graduate students in Iraqi universities are an important part of the manpower involved in scientific research, those students shape the future researchers, therefore these universities, create conditions of physical, moral, appropriate to attract. Despite the expansion of universities and increasing umbers of graduate students, the development of scientific research is still limited and attending school is in difficult circumstances.

Human manpower related to scientific research in Iraqi universities is characterized by a strong current reality with the following:

- The opportunity for holders of doctorates from the newly graduated in training and direct involvement in training is very limited.
- Scarcity of complementary research teams.
- The proportion of the number of students to the teachers of acknowledged international proportions is high.
- Preoccupation with a large number of teaching staff to work overtime.
- Opportunities for research assistants and technicians for training in developed countries are very limited.

Available statistics indicate that Iraq spent in 2003 on research and development rate of 0.4 % of GNP, and the universities have contributed hitting 31% of the volume of expenditure on research and development.

The analytical study on the budgets of university official in 2003, indicates that the expenses of scientific research (to support research projects, publications, journal and scientific conferences, books and periodicals) had reached in universities, compared with expenses of scientific research and development, in the year (2000), increased by 1.4% as it in (2000) and that most the increases were concentrated in the creators of scientific research (supporting , research projects, conferences and publishing missions and training).

The new pattern in higher education is to ensure the quality and its management, as well as that there is quality assessment and quality evaluation and quality assurance. The term assessment will receive many meanings and connotations, while the term "performance standard" refers to the level of achievements. There is confusion between standards and criteria, while using quality assessment and quality review as synonymous to evaluation.

The terms accreditation and quality assurance in education differ from one country to another. The standards are used in the USA on the same meaning of criteria, but in Europe synonymous of quality assurance; whereas the quality assurance is apart of the quality management in education.

Various concepts of quality in higher education are used to achieve accuracy and perfection through quality management. Quality is a unique kind of performance more applicable to education, as well as the quality of students, that is the ability to change constantly. Another concept is quality and the ability to report the value money and this concept has become common place, especially when something fits the quality product. There are other concepts of quality, including the concession (Excellence). Quality is suitable for the purpose of fitness as well as the traditional concept of quality of education, has been associated with the inspection and rejection. The transformation of this traditional concept of quality in education to the concept of quality assurance of education, which is based primarily on the need to test the typical rates of performance and build quality management systems for education, with the difficulties of application it appeared extremely important for the application of total quality management in education and requiring participation to ensure the survival and continuity of education institutions.

Quality education means an estimated total characteristics and advantages of the product to meet the educational requirements of students and labor markets and society and all internal and external benefit. The achievements of quality of higher education require directions of all of human resources, systems and methods processes and infrastructure to create favorable conditions for innovation and creativity.

For quality there are several concepts such as:

- Value for money.
- Added value.
- Transformation methods.
- Fitness of purpose.
- As threshold (beginning as threshold).
- Consumer satisfaction.
- Enhancement
- Improvement.

There are five areas of quality of higher education:
- A whole system of higher education.
- Foundation.
- Programs.
- Education operations.
- Outputs.

Some propose a distinction between three sets of specifications for quality:
- Scrutiny subjects.
- Measurements.

- Inferred tools.

Austin designs two criteria for defining quality, especially in higher education. <u>The first criteria</u>: is the view that the concept of quality in education must focus on the fame and reputation of the institution or its sources. <u>The second criteria</u>: is believed that the definition of quality education must be enhanced and strengthened through the application of philosophy of improving quality.

The traditional concept of quality of education is associated with the screening operations, focus on the test, then the great importance for the application of total quality management in the education to ensure the survival and continuity of higher education institutions.

Quality of education includes indicators for measuring the level of achievement in teaching and measuring of adequacy of the needs of labor markets, the number of students to each teacher, spending per students and repetition rate. It is the concept of multidimensional and dynamic levels.

Quality is important to improve the outputs of the educational process and develop the participation of the state and society, where the country has the responsibility to adopt quality standards that must be available for the department, may be unable to achieve on its own in certain situations.

Multiple concepts of quality as pointed by both Green and Harvey, including:
- Precision.
- Perfection.
- Constant change.

And Lim in 2001 put on mechanism for quality assurance in higher education that includes:
- Setting targets for education institutions.
- Compatibility with the general goals of the society.
- Test the effectiveness of quality management system to achieve the goals of education.

Quality depends on the context, which is called quality system and the message of the education and its goals. The quality varies in meaning depending on the:

- Components of education, students, the labor market.
- Frames of reference for quality, inputs, processes, outputs, and goals.
- Historical periods.

The quality of education requires the direction of all human resources, policies, systems, methods, processes and infrastructure in order to create favorable conditions for innovation of educational product.

The culture of quality and its programs lead to involvement of everyone, management scientific integrity, student and faculty member, to become a part of these programs and therefore the quality is the driving force required to push the system of education.

Total quality is the tool and process for practical application, which aims at achieving a culture of continues improvement in order to gratification, appease and please consumers and costumers, have many perceptions and concepts, including:

- Philosophy concept.
- Tools and processes.
- Continuous improvements.
- All employees.
- Gratification, appease and please of customers.

Dealing with the comprehensive quality of whole educational system requires encourage individuals to participate in decision making.

Multiple terminologies are used for classification of countries according to economic conditions, social, educational, cultural. The backward demonstrates the lack of hope in economic reform in the developing countries it refers to the poverty of performance. [The third world means the (non-developed countries), the first world (the industrialized countries of Europe and America), and the world II means (former socialists)].

The use of the entrance to the execution of quality assurance in education is an effective if certain requirements are available, such as qualified faculty members that devote themselves to work all the time at the education, available administrative services, and university leaders realize the importance and necessity of quality assurance.

The quality assurance is a process which focuses on quality measurement procedure for the institution or program and refers to a range of activities, methods, and procedures. The term is used alternately with quality management, quality assurance and quality control aimed at controlling the process and removing the causes of unacceptable performance. The overall quality assurance indicates:
- Policies and directed processes.
- Description of all the systems sources, and information of education.
- Planned and organized reconsideration of the institution or program.
- Planned and organized activities that apply the quality.
- Evaluating the ongoing process (assess, control, security, maintenance, improving the quality of education system).

The principle of quality assurance in the education contains general, polices and processes geared towards providing all help to achieve quality, and preservation they are as follows:

- The meaning of quality in education.
- Entrance performance to ensure quality.
- Quality inspection system.
- Organization of education.
- Management plans.

Quality management in education requires a system in several sequential steps serialized as follows:-
- The task of identifying the goal of the education.

- Determining the functions of the education.
- Defining the objectives of each post.
- Establishing a management system and quality assurance.
- Placing a system of quality inspection.

Defining the mission of the education must be conducted through cooperation and understanding the vast majority of the education community in accordance with laws and aspiration of state, the education of modern concepts and futurology realism of the education. The functions of the education are supposed to be modern (teaching, scientific research, community service) with clear objectives for each these posts. Teaching has goals, such as curriculum, innovative and effective with modern methods of teaching and distinguished health teaching climate. The scientific research has objectives that deal with increase of the productivity of research, publishing with marketing and modern skills. The goals of community service function are represented by accommodation of the needs of society and cooperation with other organizations of the society.

The recent trends in quality management are working to avoid a narrow view and work on measuring the output of education according to different tools that provide educational services at the level of bachelor.

Parameters of Quality Management in science

These parameters are: developments of integrated system for the education and institutions management plans, the application of the structure of modern management plans for teaching and scientific research and community service, within the clear objectives and specific strategies on several levels. These parameters require financial resources and follow, sequence in the preparation of plans at different levels of the university.

Quality management system requires several consecutive steps, including:
- Identification of the message of the education.

- Determining the functions of the education and their relative importance.
- Defining the objectives of each function of the education or college or department.
- Identification the system of quality management.
- Preparation of quality inspection system (Evaluation of the performance of the education).

The conditions for the quality assurance system include the following:
- Eligibility of faculty members with appropriate degree.
- Providing administrative services and good electronic tools.
- Capacity of education current factors in the quality of education.
- Using and adopting of scientific promotion according to academic potentials.

The inspection and screening process of quality carried out according to evaluation performance of the education and depend on the quality of the new file that contains:
- Message of the education.
- Specific objectives of teaching and research.
- Community services.
- Quality management system.
- Performance standards.

The Evaluation of education

The evaluation process is necessary as an integral part of the process of developing and colleges affiliated with. It is also necessary to identify and to achieve the objectives of the education in raising the level of scientific achievement and scientific sobriety.

Tools used to measure the advancement of the education could be summarized according to the followings:

- Measurement of the efficiency of output (graduates).

- Scientific level.
- Scientific achievements.
- Research.

The mechanisms used for evaluations depend on description and governance. Description is carried out by gathering information, whereas governance is done through combining certain characteristics that are needed often the evaluation, in addition to the identification of certain principles to be implemented, such as the purpose of the evaluation and selection tools for evaluation.

Scientific situation

The availability of financial resources in Iraq, especially before the nineties of the twentieth century has made great efforts in manufacturing, especially in the field of Military Industrialization. Iraq succeeded in building an independent military industrial base by enabling technology in modern manufacturing processes, however, strategies and manufacturing policies was clean from effective action to develp local capacity and providing appropriate incentives for local people to be able in modern industrial technology.

Despite this, Iraq still needs to pursue innovative methods to meet the daunting challenges of a large number of other sectors of production and services, as well as to develop scientific and technological capabilities of local traditional industries to modernize and to address a variety of social economic problems.

The scientific efficiency in Iraq itself has a capacity of innovative crucial role in facing challenges, in spite of standard conditions faced by the departure of many of them outside Iraq.

The limited scientific successes that have occurred in Iraq are the result of the efforts of some institutions of science and technology (Scientific

Academy, the House of Alhikma and the institutions of higher education). On the other hand, important achievements were made in building the institutions mentioned, as well as in human resources development, but the institutions mentioned and others are still far from an enabling a distinguished role in development. The linkages and synergies between the scientific and technological institutions of governmental organization and the business world remain weak, despite the existence of some versions of contracting.

The spending on research and development in Iraq, at best, less than its counterpart in the Arab countries and more discouraging regarding the outputs of science and technology, so that scientific and technological publications in specialized areas and the number of patents granted to institution and individuals are much less than the average of the corresponding figures in Other developing countries.

As a result, the status of science and technology in Iraq needs a lot of attention. The input and output of scientific and technological point indicate the deficiencies in information networks, computers, advanced equipment, scientific research.

Scientific cooperation with Arab countries, foreign states and international organizations

The current levels of scientific and technological cooperation between Iraq and Arab countries are very low, as well as with foreign countries which are almost non-existent as illustrated by the lack of joint scientific research and the outcome of joint publications. This refers to the need for Iraq in its quest to strengthen their scientific and technological interdependence.

To enhance communication and cooperation between Iraq with other Arab countries is a prerequisite in better determining the scientific problem and to obtain acceptable return of the available knowledge. Lounge of the basics of scientific cooperation is to avoid free turn key contains from any type of Technology, which is almost a priority involving scientists from two or more research fields, and incur the available possibilities to researchers to attend scientific meetings and make more use of international organizations.

The activity of transfer of science

The term technology transfer indicates that the technology is acquired through the transfer of goods or services technology and therefore is not equipped for transportation and technological deficits.

The experience of Iraq in the transfer of technology is varied and appropriate in a large proportion, completed with foreign companies which provided comprehensive and complex technological deals in the framework of international market strategy, and the country suffered from indiscrimination of transfer that took place in the absence of a sound domestic policy in various technological fields.

There are two problems facing Iraq, the first is concerning the search, transmission, absorption, development and improvement of modern technology, and second is related to technology development. According to that, Iraq needs in the area of technology transfer the following:

- The search for technological alternatives.
- The selection of appropriate technology.
- Adapting the selected technology.
- Identifying the problems of adapting modern technology.

Research and development "R & D" activity

The research and development are both vital and active in maintaining the quality of scientific personnel, in ensuring access to advanced science and promoting technology transfer. In addition it is provide early warning preparation of technological progress, industrial, agricultural and health, whereas its investment is guaranteed.

The efforts in research and development in Iraq may differ from what made similar efforts in other countries in terms of outcome, it is still inadequate to meet the challenges posed by scientific and technological developments and the process of globalization despite the allocation of funds required in this area, and the last twenty years is good evidence.

- Establishment of an information society.
- Innovation of small and medium-sized enterprises participation.
- Development of human potential through the training of researchers.

Science policies in Iraq (Higher Education and other sectors)

There are no clear scientific and technology policies in Iraq, but may be the presence of strategies development that include the lines of long-term development of science and technology indirectly that deals independently within the development process. In Iraq private enterprises will invest in science and technology (Higher Education .... and others) in establishing a scientific and technological infrastructure.

The scientific, international cooperation in Iraq is limited and weak at present, while the efficiencies gained by scientific institutions in Iraq in coordination with existing competencies in the areas of social economic activity exists in society.

Under scientific siege, budgets reduced drastically, and scientific institutions reduced also their expenses strongly which led to the non continuation of the previous level of innovative capacity, and the inability of the infrastructure and science and technology to do their core functions.

Iraq continued to practice, so far away from any genuine interest, the development of science and technology policy, due to many reasons already dealt with. It is worth mentioning here that any scientific and technology policies, in particular, depend on the quality of Iraq's exports. Petroleum Exporting Countries and Iraq, supplier of natural non-renewable sources reflected the policies of science and accordingly, it requires the decision-making process and scientific identification of reliable channels between scientific community and the political leadership. Moreover, the wording of policy drafting requires determining the objective of science and technology that includes:

- Promotion the growth of local companies specializing in services and manufacturing activities.
- Advancement of specialized institution.
- Increasing the movement of personnel in science and technology.
- Overlapping science policy and technology with a range of social and other economic activities.
- Development of institutions that will depend on science and technology that operates in a different environment.
- Strengthening the management structure of science and technology.
- Raising the level of resources.
- The effectiveness of R & D institutes.
- Development of technology transfer.
- Development of international cooperation.
- Improvement of researchers' wages and working conditions.
- The development of science and technology in long- term visions.
- The establishment needed in long term visions of all relevant institutions.
- Definition of the roles of effective government, private sector institutions, non- governmental organizations and professional associations.
- Adoption of progressive methods to build scientific capacity.
- Removal of duplication, overlapping and conflict in science organizations.
- Monitoring and evaluation and effective drafting in the relevant agencies in science and technology policies

It is proposed that a variety of science organizations should be related to the formulation of science and technology policies, including scientific societies and other professional such as industrial and technological organizations.

These organizations have to operate in an environment different from the classical environments in recent years (currently confined to the production, dissemination and application of knowledge covering a range of disciplines and areas of application), but must evolve and interfere with a range of social and economic activities and cooperate with other private sector organization. Moreover, these organizations propose concepts and scientific policy inputs.

Moreover, these scientific organization should involve the largest possible number of workers participates in the preparation of science and technology policies, including representatives of government departments concerned economists, chambers of agriculture, industry, trade, non-governmental organizations and scientific societies.

Research institutions and administrative development

The efforts in research and development in Iraq may differ from what efforts made in other countries in terms of outcome and the challenges posed by scientific and technological developments and globalization. The linkages between research and development do not meet the need, as they are usually not based on cooperation, and are sometimes contradictory because of the lack of flexibility and bureaucratic practices.

The problems on research and development requires from its institutions to evolve, according to the ideas that have been developed. These relate to the maintenance of the quality of special scientific cadre. Further more then to access to renewable science in the world, promoting industrial development, technology transfer, contributing to social, and economic planning and accordingly proposed that the research and development institutions have special functions as well as:

- Proposals should be submitted through the channels when developing the new science policy.

354

- The possibility of providing early warning systems in preparation for preparation for technological change.

Scientific Academy

It is proposed in this area, that the scientific academy have a complex role in the development of science policy and is considered to be one of the main tributary of the scientific policy and may be a substitute for the supreme for Science Technology. The participation of specialized committees for example, could provide complex scientific projects, relating to environmental pollution, energy and natural resources projects and areas of agricultural development, public health and natural resources.

The specialized scientific societies could propose several projects and comprehensive survey of scientific resources, human and various national materials. Also enhance the working groups that can be developed and networks of cooperation between scientists from various disciplines concerned, with the establishment of networks of information for selected sectors based on the priorities identified by policy. It can evaluate research projects and public awareness of science and technology and increased attention to scientific research.

Industrial enterprise and science

These institutions are required to have the presence of a specific set of targets linked with science policy and technology, which are consistent with a coherent vision for the future.

The responsibilities to the views of specialized advisory functions include characterization and strengthening the linkages and knowledge flows and providing technical services.

Iraq's various institutions face difficulties in promoting scientific and technological capabilities and future planning of the scientific, issues to acquire the capacity of innovation. This can not be achieved unless being done within the framework of science policy interrelated. Accordingly, it is required to use new methods of science and technology policies, and to develop integrated growth strategies that take into account of local and external scientific conditions. To implement these, it is proposed to develop an academy or a higher council for science and technology. This Council (or academy) which is an institution with personal, moral, financial and administrative requirements characterized by autonomy should be linked to the Office of the Presidency.

The council carries out the followings:

-       Development of science and technology policies in accordance with the demands of current and future science.

-       Contributing to the scientific development of internal and external sources.

•   Promotion of scientific studies and research in the country to keep abreast of scientific advances in the world.

•   Establishment of scientific ties and close cooperation with Arab and international destinations.

•

The academy or the Supreme Council consists of:

-       Members of not less than 35 not more than 40 including the President of the Council or the Academy.

-       Secretary-General.

-       The member should be a scientist and researcher in one of the branches of knowledge (agricultural, industrial, treatments, medical, engineering) and has a broad access to one or another branch of knowledge, and has genuine scientific product.

-       The academy members should be appointed by the Prime Minister and enjoys the rank of Deputy Minister.

-       The Academy full-time secretary-general is appointed from among its members.

- The Academy has a number multiple specialized committees working in coordination with other scientific institutions in the country (Commission on Technology, the Committee of Water, Energy Commission and Commission for information).

- The Academy has specialized offices within the framework of scientific knowledge (agricultural, medical, engineering, pure science).

- The Academy creates the suggestion of science and technology policies and then filed for the presidency for the Adoption (after being revised, by its committees and services).

- It also works with the recommendation to grant material assistance to the centers and individuals, adoption of the establishment of various scientific centers.

- Development of labor contexts between scientific institutions and the beneficiaries and the mechanism of cooperation between scientific institutions of Iraqi and international institutions involved.

Proposing scientific policies

The Supreme Council for Science and Technology proposes a policy for science and technology in the country in accordance with the following tasks:

- Coordination of the efforts to formulate science and technology policies in Iraq at the institutional level.

- Clarification of methodologies used in policy formulation of science and technology.

- Modernization and integration of science policy with development policy.

- Analysis of scientific and technological capabilities in Iraq with a focus on research and development.

- The need to inform policy makers of the methods used in the planning and programming of research activities

- Building local capacity through education and experimentation.

- Following up the implementation of these policies through the secretariat and a mechanism to provide on issues.

Formation of a Higher Council for Science

Detailed proposal:

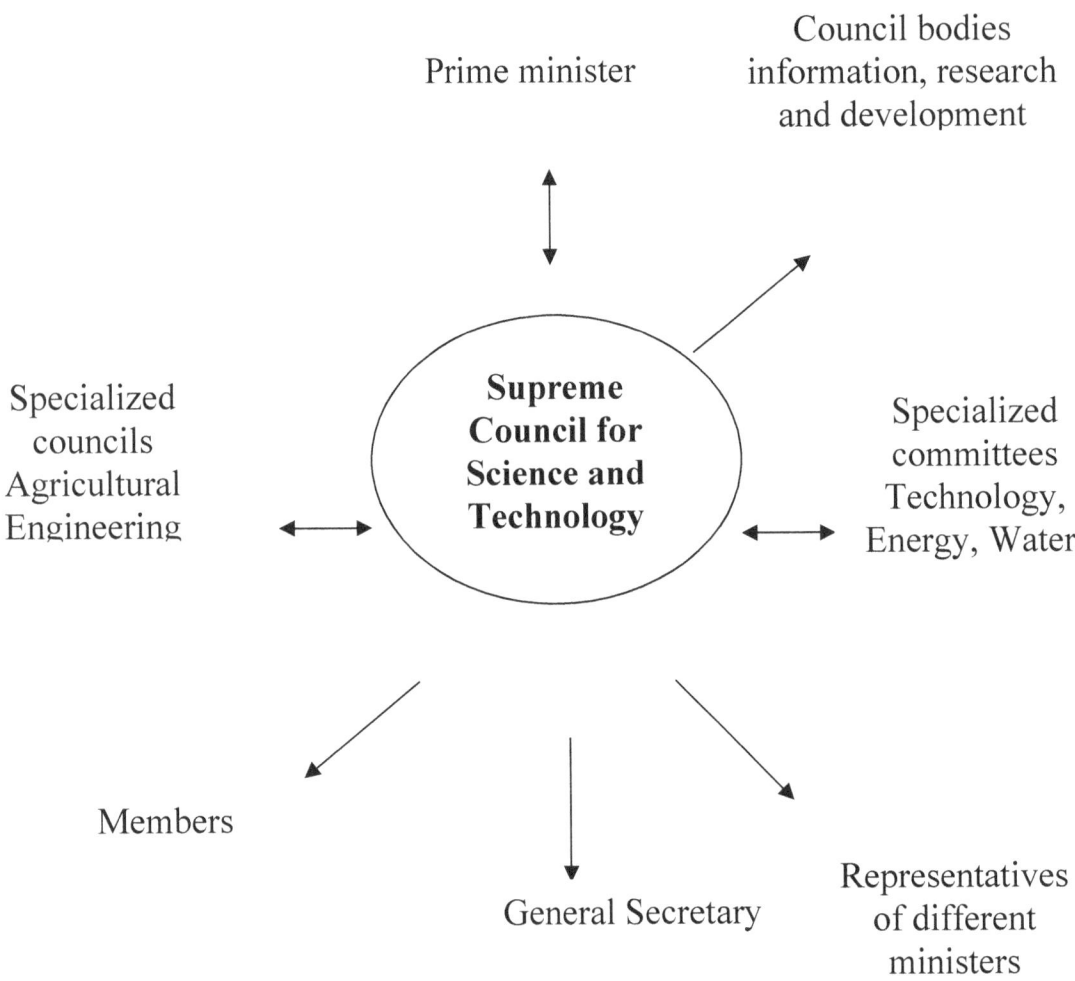

Towards formulating policies of science for the twenty first-century

The creation of the council facilitates and exercises science and technology policies, including matters related to the twenty-first century in Iraq. The need is to create some things that have been reviewed in the preceding paragraphs, including a brief analysis of scientific and technological capabilities, with a focus on research and development as well as the change that happened in scientific and technological institutions, leading to social and economic development balanced with the following characteristics:

- The scientific development and technology should be based on the local capacity and the scientific and technological progress made in other countries, with balance between the two sides.
- Investment of scientific institutions of science and technology in creating the scientific and technological infrastructure.
- International scientific cooperation.
- Promotion of development of national technological capabilities in both the private and public alike.
- Coordination of competencies gained by scientific and technological system and their integration with national huge number of competencies in the areas of social and economic activity.
- The integration of science and technology policies with a set of national policies and programmers that meet this purpose.
- Continuation of production and dissemination of scientific and technological knowledge and maintain an effective level of innovative capacity.
- Continuous documentation of formulating policies and scientific technology.

Policies can be characterized by scientific and technological trends and are generally maintaining the balance between any movement toward the development of local scientific and technology and the establishment of links with external sources of technological knowledge that take place in several stages:
- Continuous assessment of the current status of science and technology.
- Development of the prospects of continuing to develop science and technology during any (year).
- Development of a strategy based on specialized studies related to the objectives of developmental policy.

- The preparation of a detailed five-year plan accompanying executive programs constitute the followings:

- Keeping pace with scientific and technological developments in the world (biotechnology, information technology, energy technology… etc).

- Keeping up with scientific and technological developments in the world through the follow-up scientific and technological capabilities such as (biotechnology, information and energy).

- Monitoring the change in scientific institutions and the Arab world and the extent of the involvement of scientists in Iraq with their peers in the Arab countries and the world.

- Following the changes in social and economic conditions throughout the world.

- Examination of human skills constantly has been much more important than raw materials and it leads to many possibilities, including:

- Achieving parity with other countries in science and technology.

- Identifying better scientific problems and understanding of the issues that are raised in a more comprehensive way.

Development of the science of biotechnology

Iraq occupies a very diversified geographic and climate zone and assumes a great deal of importance. However, its wealth is under U.N. resolutions. Its health and agricultural problems are diverse.

For almost twenty years (1970-1990), its Biotechnological activities have sought to bear on various problems on Agriculture and Medicine. In recent years (1990s), these activities have turned in a very limited way, to use some of new developments in Basic Biotechnology such as Molecular Biology and Genetic Engineering.

Various Centers and Departments are involved in the National Biotechnology research systems in Iraq. The Biological and Agricultural research Centers also integrate such activities into their research program.

Presently, these centers established different types of programs (long term, short term) in Agriculture and Medicine, for generation, development and transfer of Biotechnology. Major components of these programs; include development of technology transfer facilities and establishment of international collaboration in the field of Biotechnology.

Medical Biotechnology in Iraq, as a whole, demands strong collaboration in order to reduce the present gap between Iraq and the developed countries. Whereas, Agricultural Biotechnology seeks to increase food crop resistance to insects and diseases, utilize a wide varieties of techniques, including Biological control and various forms of Biotechnology, such as tissue cultures, molecular genetics and genetic engineering.

This part of the analyses of the major activities of Biotechnology in health, Agriculture and Basic Science in Iraq offers general background information on the development of Biotechnology. Then examines the current of Biotechnology research and potential, and suggests an outline of the strategies for the future of Biotechnology in Iraq.

Iraq, which has already entered the biotechnology era, whether on purpose or by chance, will in any case inevitably suffer the consequences of the widespread use of such techniques in industrialized countries. Therefore, the question is no longer whether such a move is desirable. but more how can best be put into use.

Currently people of Iraq are suffering from poor quality of life because of U.N. sanctions, which resulted in:

- Food shortage and malnutrition.
- Unsafe drinking water.
- Improper sanitation system.
- Poor health care.

This is why Iraq needs to have at least a basic level of skills to make it possible to define and implement policies in Biotechnology and the other technologies. Biotechnology alone will not feed Iraqis or give them better health, but it would be irresponsible not to see it as one of the available tools. A certain number of social, cultural or institutional conditions also have to be satisfied before technical success can be converted into economic and social progress.

This part analyses the major activities of biotechnology in Iraq; then examines the prospects of strengthening biotechnology in conjunction with conventional technologies; and finally it outlines the strategies for the promotion of biotechnology in Iraq.

Approaches of biotechnology in Iraq

Most biotechnological activities, which are applied in Iraq, are limited to traditional methods and serve their needs through:

- Fermentation
- Antibiotic industry
- Single cell protein
- Plant biotechnology
  - Tissue culture
  - Soil fertility, through biological activities
  - Increasing food production through plant cell-culture
  - Bioconversion of waste for food and feed ingredients

Fermentation

Iraq in early 1970s established traditional fermentation industries, bakers yeast production, ethanol, acetic acid, acetone, butanol and citric acid production. In 1970, a factory for making bakers yeast from sugar-beet molasses was established with plans for the production of compressed yeast.

This industry was faced by many problems, especially in dried yeast production:

- There are abundant sources of raw materials for fermentation in Iraq. Large quantities of hydrocarbons, and carbohydrate by-products (molasses) and lignocelluloses waste are found.

362

- Fermentation of food crops like dates, which enjoy comparatively large market sizes, does not receive sufficient attention. Local research on bioreactors is therefore needed to support the development of new processes and improve the performance of food industry in Iraq. Bioreactors also play a key role in the production of enzymes used in the beverage, detergent and leather industries.

Antibiotics industry

The antibiotics industry in Iraq started in 1970 for the production of penicillin and tetracycline. On the other hand, tetracycline production continued until 1980. This industry was discontinued for economical and technical reasons as a result of sanctions. The involvements of research and development programs have positive effect for restarting bio-industry and production of different types of antibiotics.

As the pharmaceutical industry in Iraq is directed to satisfy the local market by satisfying the market needs, it is therefore necessary to develop appropriate biotechnologies against various diseases that are endemic.

Single-Cell protein

A research and development program for single-cell protein (SCP) production at pilot plant level started in Iraq in 1982. The production of SCP from methanol using local and imported strains of Candida Utilis was investigated. The research plan included an economic feasibility study and the assessment of technological and nutritional aspects of SCP under local conditions. Then, another pilot plant was established to utilize date syrup for the production of bakers yeast and SCP.

The most important achievements of this program were the establishment of pilot research facilities, training of personnel, the nutritional assessment of available commercial SCP products and the isolation of several methanol-utilizing bacterial cultures.

Plant biotechnology

Plant biotechnology applications in Iraq include soil fertility (nitrogen fixation) using yeast strains in a mixed culture and cell conversion. Scientists have carried out research on nitrogen fixation by grain legumes. The main objective of this research was to increase the yields of grain legumes while decreasing the input of inorganic nitrogenous fertilizers

Tissue culture

- In view of the importance of the date-palms in Iraq, a tissue culture laboratory was established in 1979 at the Agriculture and Water Resources Research Center in Baghdad. Another laboratory was initiated at the Genetic Research and Biotechnology Scientific Research Center for the improvement of plant production. Tissue culture laboratories were also established at the Universities of Basra and Musol and also at the Ministry of Agriculture. The latter worked in collaboration with FAQ through regional center for date-palms in the near east and North Africa. The major objective was the commercial propagation of plants by in vitro techniques.

- Then in 1982, the vegetative micro propagation through tissue culture was carried out as a promising technique. However, future research is needed to early following and lack of uniformity of the closed plant.

- Date palm propagation by tissue culture was also implemented at research institutes and universities in Baghdad and Basra. Other species such as lettuce and potatoes were propagated by tissue culture at the department of Biology of Mosul University.

Bio-control programs used for controlling pests

The plant production research center is a State Board for agricultural research. College of Agriculture, Baghdad and Mosul Universities,

Agriculture and Biology Research Centers and Iraq Atomic Organization are the main research centers in Iraq, which carried out different research studies on the biology, taxonomy and control of the pests. These research centers successfully adopted control measures on the most important pests attacking crops, vegetables and by applying chemical and agricultural methods. Promising results were obtained from many plant extracts against pests, by inhibition pest life cycle.

In Iraq, most of the agricultural land subjected to grow many of weeds causing big losses to the agricultural crops, as many research workers confirmed the positive results of the weed control measures to increase the yield of the crops.

Researchers started to evaluate herbicides since 1965, and mid of seventies, herbicides were applied to control the weeds in wheat, rice, corn, cotton, potatoes and tomatoes. Increased support is needed to expand research designed to develop new herbicides that are not likely to pollute ground water and that will provide reliable control.

Several centers are involved in bio-control programs such as:

Commercially, research centers introduced two bio-control mutant fungi. Both fungi successfully were applied to control plant parasitic nematodes and soil born fungi on vegetables and citrus. Also the center used another fungus against date-palm stern borer insects.

State Board for agricultural research successfully adopted several control measures on the most important pests attacking field crops, vegetable and fruit trees by applying chemical, biological and agricultural methods. Many insect growth regulators, bio-control agents, fungi and plant extracts were experimentally applied on small and large scale fields. Most of the research studies of the graduate students in the department of plant protection concentrated on the biological control, and plant extracts.

At the present time, the U.N. sanctions which have been imposed on Iraq since 1990, destroyed most of biological control programs and Iraq is facing lack of well trained personnel and shortage in facilities. As a result the first generation biotechnologies used in Iraq such as insect resistance, herbicides resistance are not easy to address any more.

Health biotechnology

- Several centers are interested in carrying on research on health biotechnology. The following may be mentioned.
  - Previously Saddam Center for Cancer and Medical genetics Research (SCCMGR).
  - Institute of Biotechnology and Genetic Engineering for Graduate Students, University of Baghdad.
  - Department of Genetic Engineering, College of Science, University of Baghdad.
  - Genetic Engineering Departments in a number of Universities.
- SCCMGR is engaged in several lines of biotechnological activities in health:
  - Cloning of tetanus toxic gene into tumor cells.
  - Preparation of tumor cell lines in vitro for gene therapy technique
  - Preparation of restriction enzymes vectors for gene therapy.
  - Studies on disorders of mitochondrial DNA in muscular dystrophy.
- Several biochemists have participated in various research projects that deal with diagnosis, and monitoring of several types of tumors.

Health and medical technologies

- Gene therapy and diseases
- Markers of neoplasm
- Genetical markers
- The production of vaccines
- The production of antibiotics
- The production of hormones and enzymes
- Vaccine production
- The production of antibodies to microorganism and toxins from snake bites and insect bites

- The production of antibodies (monoclonal antibodies)
- Development or establishment of cancer lines in the laboratory
- Transfer of bone marrow and cultivation of bone marrow
- Using molecular indicators in human lymphatic cells
- Diseases of hereditary Cancer
- Diagnostic kits
- Forensic-genetic fingerprint

Environmental Bio- technologies

- The Use of living organisms in purification of heavy metals of the environment.
- Preparation biocides to combat agricultural pests.
- Preparation of bio-fertilizer to improve agricultural product.
- Finding microorganisms to revive the disintegration of some hydrocarbons and turned into a simple compound.
- Bio-technologies in the ecological balance.
- Bio-treatment of sewage.
- The use of microbial treatment to remove oil, chromium and ammonia.
- Using natural organisms to convert hazardous substances in the oil.
- Clearing industrial sites and avoid pollution.
- Oil Pollution by breaking chemical compounds.
- Isolation of bacterial strains that have the ability to remove sulfur.
- Using Bio-Markers and Bio-controls for detecting pollution levels.

Bio-technologies of Water

- Aquaculture techniques
- Techniques of bio- actors of water
- The use of bio- sensitivity
- The use of drugs and vaccines to preserve the health of fish.
- The cultivation of fish.

Biotechnology and basic sciences

- Genetic engineering techniques
- Genetic fingerprint (D.N.A. Finger Printing)
- Production of restriction Enzymes
- Production of standard D.N.A.
- Diagnosing the production of certain genes
- Amplification of genetic materials

genome (genetic content)

- Studying the genetic content of the bacterial isolates
- Isolating and purifying D. N. A.
- Determining the content of the Plasmid isolates
- The safety of animal products and genetically modified organisms
- Building gene maps of plants of economic importance

Studies related to bio-technologies

- The systems and regulations for genetically modified animals
- The importation of transgenic animals
- Economic feasibility of genetic modification in animals farm to produce therapeutic human proteins
- Controls systems in the use of genetically modified organisms
- Ethics and social values and bio- technologies
- Bio- informatics
- The Bio-safety and bio-technologies
- Laws regulating the use of genetically modified crops
- Medical, religious and security considerations of bio- technologies

Scientific heritage

Theories in circulation and the most specialized in the formation of life:

- The fall of some organic molecules on earth from comets.
- General fertilization year (planting the ground by life from the intelligent beings inhabiting advanced planets).
- The emergence of life after eons of consecutive chemical filtration in the sea.
- The emergence of life quickly, after the land formation in brief time.
- The emergence of life through escalation of dusty water droplets to the surface, the collected chemicals turned quickly to life.

Accordingly these theories can be divided into those and other to:

- Theories of life

Include the notion of self- evolution of life, this theory has emerged when ancient Egyptians believed that frogs emerged from the mud of the Nile and the old Greeks put forward the idea that rats originated from the trash, and renewed self- intestinal parasite. Then, the theory stated that living organisms reproduce itself from living organisms supported by Louis Pasteur (1872).Then ,the idea of self- evolution, stated that living organisms can not arise unless from similar living organisms. Accordingly, the self- regeneration, was collapsed which was mental prevalent, depends on the physical, through science, including the development of the first microscope before, the microscope has been able to prove that the drop of water (rain) where there are no bacteria, but generated after dropping to the ground.

Models of inherited science

Scientific heritage is the most precious legacy of the human to human being and the complete presence of human that provide him with the meanings that make sense for a human to distinguish them from other organisms, however at the Arab and Islamic library of hundreds of men did not lift them up after the dust of oblivion, .

Methods and ways of scientific research in the scientific heritage.

The event in 1996 was exciting, the birth of Dolly in a somatic cell into a specialized egg-enriched after removing core and planting it in the womb; the most important point of this event is the return of specialized cell and embryonic stable situation after losing this status. The other development in science is the production of sufficient quantities of food in the world. Many thinkers expect that the world will see a lot of problems, related to scarcity of resources and energy, such as increased pollution and population explosion. Most studies of future ending 2025 required further means such as:

- Technical means (computers to store and recall information).
- The use of special programs to make the prediction.
- Many experts, technicians and programmers.

Possible divisions of future studies in science include three types; this division is used haphazard to simplify them, as it is difficult to separate the three types of studies, from each other, such as:

• Studies that rely on prediction (what will happen in certain area of science).

• The overall outlook studies that rely on intuition which looks at the impact of current scientific achievements on the future of humanity.

• The overall outlook studies relied on detailed statistical information within the mathematicians programs on computer models.

.

There are several theories to explain the emergence of life and trying to answer the classic question. How life was began? The answers such as; the fall of some organic molecules on Earth from comets and planting the land with life by the intelligent beings on advanced planets. The emergence of life after eons of consecutive nomination (years) with quick chemical composition following the ground composition .Then after the short time period (the escalation of water droplets to the surface followed by collected chemicals, turned quickly to life in other place, then moved to the ground).

After the general review of ideas about the evolution of life, the question was raised about the secret of inductive creation of the universe and the concordance between the creation of the universe and life.

There are clear trends in the study of the secret of creation of the universe and life, firstly believe that there is an aim in it and secondly the deny of the existence of the end, but to the existence of chance. Among the topics that are interesting mystery of the creation of the universe and of life, origin of the universe, evolution and various theories that talk about evolution, which accept both science and reason, scientists and philosophers. As well as the emergence of life and theories of ancient and modern, material and ideal, west and east that deal with them. The evolution has contributed to clarify the continuing march of life regardless of the theories put forward, including Lamarck and Darwin.

From an examination of creation as described in the Qur'an, an extremely important general concept emerges, it is possible to compare the six days of creation .

"Your Lord is God who created the heavens and the earth in six days." Qur'an, 7:54

. It refers both to the heavens before the earth and the earth before the heavens, when it talks of creation in general, as in this verse of chapter Taa Haa:

"(God) who created the earth and heavens above." Qur'an, 20:4

These ideas are expressed in chapters Fussilat and al-Anbiyaa:

"God then rose turning towards the heaven when it was smoke" Qur'an, 41:11

According to modern science, the separation process which appears dozens of times in the Qur'an. For example, look at the first chapter of the Qur'an, al-Faatihah:("Praise be to God, the Lord of the Worlds." Qur'an, 1:1 These Qur'anic references are in perfect agreement with modern ideas on the existence of primary nebula (galactic dust), followed by the separation of the elements which resulted in the formation of galaxies and then stars from which the planets were born.

. Theories of the creation and formation of the universe

Expansion of the Universe

Chapter ath-Thaariyaat of the Qur'an also seems to allude to one of the most imposing discoveries of modern science, the expansion of the Universe.

"I built the heaven with power and it is I, who am expanding it." Qur'an,51:47 The expansion of the universe was first suggested by the general theory of relativity and is supported by the calculations of astrophysics.

Authority to travel in space can only come from the Creator of the laws which govern movement and space. The whole of this Qur'anic chapter invites humankind to recognize God's beneficence. The universe consists of hundreds of billions of galaxies, each with hundreds of billions of stars. This myth of seven heavens was a common idea prevalent during the time when the Qur'an was first recited. See ye not how Allah has created the seven heavens one above another.Qur'an 71:15

Surely We have adorned the nearest heaven with an adornment, the starsQur'an 37:6

The earth first formed around 9 billion years after the Big Bang. The Qur'an, however, repeats the prevailing middle eastern myth that the earth and universe were formed in six days. We created the heavens and the earth and all between them in Six Days, nor did any sense of weariness touch UsQur'an 50:38

Modern science has proposed that all the elements that make up the earth (Oxygen, Nitrogen, Carbon, Iron, etc.) was originally formed by nucleosynthesis in stars and then expelled into the universe when those stars supernova. He it is Who created for you all that is in the earth. Then turned He to the heaven, and fashioned it as seven heavens. And He is knower of all things. Qur'an 2:29 . But according to this theory, the Universe was formed about 13.8 billion years ago due to arapid expansion from singularity. The earth was formed 4.54 billion years ago from accretion of debris that surrounded the precursor of the Sun.

Many theories put forward on the creation of the heavens and the earth that were formed from the old material was scattered in space and the earth were not present before. These theories are summarized according to the following:

Intellectual life of human beings

The scientific progress made by man is in fact the result of intellectual growth in the presence and the disclosure of the laws governing the universe after the dovetails of the theoretical thought with that at experimental. In the modern era we find that the traditional dividision between the old mental philosophy, which assumes that the mind is the source of knowledge and the experimental philosophy which considered the experiement is the source of knowledge is still not significant fading. However, we believe that the following ideas may help in determining the intellectual life of human beings:

- The reason is a large scale in human thinking.
- The experiment is an important tool for the application of standard mental, but not alone.
- The trend is assumed that the mental does not ignore the role of experience in science and human knowledge.

Education and scientific heritage

Each community the public and the private has a scientific legacy, with various pillars such as cultural, social and preserving the heritage is a national and human necessity. The scientific education can contribute to the benefit of the scientific heritage. So it is really important to stop a bit to show how to deal systematically with this legacy, thereby the process of transforming extrapolation of tradition and its benefits to the renewable power of the development of general education. There are methods and appropriate methodologies used for the purpose of benefiting from inherited science, including:

- Scientific heritage is treated with inherited methodology based away from the indiscrimination and excessive, damaging to these inherited disciplines.

- Scientific heritage depends on inherited as a single bloc in terms of time and the researcher does not fall blackouts, at the wrong focus on specific topics.

- The researcher shall comply with the spirit of scientific criticism of the inherited and not influenced by intolerance and prejudice.

- Selection process in the inherited case depends on modern scientific techniques.

- The researchers, scientists and leaders do not deny the human rights of those discoveries which have enriched as:

  - The invention of writing by the Egyptians and Sumerians.
  - The invention of printing in 1456 by the Germans.
  - The invention of the computer by the British and the Americans.

  - The invention of the information revolution by Bill Gates.

Profiles of the scientific legacy

The scientific heritage of Arabs lost landmarks such as the incentives to neurological and careless ignorance and the apathy of intent, and scientific heritage that reflects the achievements of the contemporary as well.

The inherited problems in science need to be solved this can be achieved only through a profound understanding of style and a modern scientific approach, according to the following considerations:

- The legacy of the scientific aspects of the spiritual and intellectual heritage is the least fortunate in the detection and study.

- The inherited issues may reflect our presence, in the most
-

appropriate and worthwhile study, but being be more careful and conscious comparison.

- The trends and aspects of heritage should and must be taken interrelated and not separate from each other.

- The texts and documents must make sure that does not carry more than they can handle and does not indulge their dangerous ideas to interpret the mandate of the planned results in the history of civilizations.

- The task of finding of scientific heritage requires keeping in mind that what unites the science is not the subjects but the method of scientific activity. In essence a systematic activity plan follows the scientific course and not accumulation of facts and events.

- The impact of religious belief should not overlook in the scientific atmosphere and do not look to science separated from the religious atmosphere completely.

- That modern science is not a specified range of facts and not humanitarian effort called for substantial probability and likely to happen.

- Arab scientific heritage includes all kinds of science and knowledge, including pure science, mathematics, natural sciences and biotechnology.

- The high percentage of scientific heritage to the Arabs do not deny that sharing of the non-Arab such as Persians, Turks, Indians and Chaldeans and other species that have entered into Islam, but all were for all their contributions to the development of Arab-Islamic scientific movement.

Scientific research in the scientific heritage

Zeki Naguib Mahmoud, the Egyptian scholar, stated that the doctrine

of Jabir Ibn Hayyan carried out in the various steps of scientific research 26are similar to the steps agreed upon by most of the scientific methods applied, which are summarized in the three following main steps:

- The scientist must be inspired by what he took at the world, presumably posed to explain the apparent phenomenon to be interpreted.

- From this hypothesis it is possible to deduce consequences of the interface by the theory of exchange.

- These results could be explained from the nature to see how can the person believes in it or not, specifically his observations.

Jabir in Hayyan established experimental methodology which tells us about training.

Geber in his experiments used in the measurement of the absent witness in the area of his chemical research, which meets the experimental method in the idea of probability and it is not permissible to judge what has not been seen. Jabir succeeded in crossing the magic chemistry to the fabulous scientific chemistry with experimental approach and relies on observation and experience.

The experimental approach of was started by Jabir Ibn Hayyan and although he was one of the geniuses who worked in philosophy, logic, observation, mathematics, chemistry, mechanics and astronomy, but he was overshadowed by the fame of chemistry and known as the nations of this art is indisputable.

The Arabs and the Muslims in the use of the scientific method according to now-a-days have the credit to advocate the use of experimental method in the study of natural phenomena and biological molecules and on an extrapolation of these phenomena and observation of the experiment, testing and surveillance.

The scientific method of research that applied by Arab and Muslims scientists like Hassan Ibn Al- Haytham and Jabir Ibn Hayyan, Jabir al-Battani, Hamid bin Musa Khwarizmi and Abu Bakr Muhammad Ibn Zachariah and Elmejriti and abu Rayhan al- Bairouni, Ibn al- Nafees and other with more accuracy than the cultural approach applied in our time.

Chapter nine

Concepts of science

ألعلم مفهوم ومنهج

أن النكبات التي تواجه الأمم ومنها التخلف في قطاعات الدول المختلفة تشير الى الضعف على المستوى العلمي لان النظام الدولي يعتمد على استثمار العلم وليس الى سوء استخدامه وأن ما يجب الوصول اليه خلق نظام دولي جديد يركز على المشاكل العالمية الحالية والمستقبلية ويأخذ بنظر الاعتبار التخلف الذي تعانيه حوالي ثلاثة أرباع البشرية وبالتالي يتطلب توفير القاعدة الفكرية ودراسة تحديات الثقافة واستيعاب مفاهيم العلم ودورها في التنمية الاقتصادية ودراسة وتحليل التخلف واستيعابه في منطقتنا العربية، ومعرفة العلاقة بين العلم والتقانة وتطورها عبر التاريخ.

أن التقانة عموماً سبقت العلم وبعدها بدأت تعتمد أكثر فأكثر على انجازات العلم، وان العلم المكتشف شخصية مميزة لها مجالات عدة لتوضيح عملية الاكتشاف والاختراع ذاتها والتي في ضوءها يمكن فهم دور التقدم التقاني في عملية النمو والتنمية وفهم عملية البحث والتطوير لمتطلباتها، والتي تساهم في بناء مفهوم للتقانة ودور الشركات المتعددة الجنسية في حالة عدم ملائمة التقانة .

أن الاعتراف بأهمية العلم معروفة تاريخية من خلال الحضارات المختلفة (الفرعونية والسومرية وغيرها)، مروراً بالجهود التي بذلها محمد علي في مصر [1]ودراسة شبلي الشميل (1853-1917) المتمثلة بنظرية النشوء والتطور وبيولوجيا الخلية ، الذي اعتقد في عام 1896 في كتابه المرسل الى السلطان عبد الحميد، حلل فيه أسباب ضعف الإمبراطورية العثمانية، التي اعتقد ان العلم احد أسبابها.

أن العلم نشاط ثقافي يحقق أهداف المجتمع الاقتصادية والسياسية ووفق ذلك فالعلم والتقانة قوتان رئيسيتان تدعمان التغير التاريخي والاجتماعي والوطني والدولي، تتباين من مجتمع الى آخر ,أما التحديات الداخلية والخارجية فتختلف من ثقافة الى أخرى، فالشرق

378

الأوسط منطقة تتقاطع فيه ثلاث قارات آسيا وأفريقيا وأوربا حيث يمثل هذا الموقع وعلى سلطانه فوائد ومساوئ.

بدأ البحث في العلوم مع بداية الانسان نفسه حيث كانت تنتقل المعارف العلمية البدائية من جيل الى اخر عن طريق المهن، وخلال نهايات القرن التاسع عشر كان يغلب على المعرفة العلمية الطابع الديني والوصفي وتركزت في الكنيسة. اما المعرفة العلمية في بدايات القرن العشرين فكانت تتميز بزيادة اقبال ذوي مستوى اجتماعي واقتصادي متواضع على الدراسة العلمية.

اما خلال منتصف القرن العشرين، فقد تم التركيز على اثراء الحياة بالبحث العلمي ووفق حاجات المجتمع ودور العلم كقوة اجتماعية وتطورت اثر ذلك مناهج العلوم وظهرت حركة اصلاح في السبعينات تدعو الى ضرورة تطور المناهج للابداع العلمي , وفي عام (1983) صدر تقرير بعنوان امة في خطر في الولايات المتحدة الامريكية وهي اشارة تحذير للمجتمع.

ان تعليم العلوم ينبغي ان يتوفر فيه مبادىء وشروط ومنها ان يكون التعلم مبكراً للطفل وفي مرحلة الابتدائية والاعدادية وتطوير القدرة على استخدام الاسلوب العلمي لحل المشكلات واكتساب الايجابيات العلمية وان المعرفة العلمية يجب ان يكون لها دور في تحقيق اكتساب الحقائق بثورة وظيفية واكتساب المفاهيم للمبادىء العلمية واكتساب المهارات لحل المشكلات وتنمية الاتجاهات العلمية.

ونظراً لكون المعرفة العلمية يقصد بها تنمية وتطوير قدرات ومهارات الافراد من اجل مواجهة متطلبات الحياة باوجهها المختلفة، وكذلك المجتمع. وعليه فان التنمية وتطوير قدرات مهارات الافراد هي بناء الانسان .

يلعب الانسان الدور المميز في المعرفة كون الاول نشاط انساني والثاني مكون انساني وبنية اجتماعية والعلمية عموماً تلازم الانسان طوال حياته. وبالنسبة للحياة فالانسان كائن حي يمتلك بناء جسمي محدد مترابط مع حالته النفسية والوظيفية.

وبالرغم من هذه التصورات فالانسان يبني تصورات مختلفة في الفلسفات العامة فهو عند الفلسفة المثالية كائن روحي يمارس الارادة ومسؤول عن تصرفاته. ويذكر افلاطون ان الانسان مؤلف من جزئين احدها ينتسب الى عالم المثل (النفس) والاخر لعالم الحس (البدن) واكد على فكرة الثنائية المذكورة بينما يرى الواقعيون ان الانسان كيان عضوي وميول اجتماعية والفلسفة الطبيعية تشير الى ان النفس الانسانية خيرة وان الحاضر اصل تطور المستقبل والتربية هي ضرورية ما دامت تستمر طول العمر والفلسفة الاسلامية ترابط ابعاد الطبيعة الانسانية (الجسم، العقل، الروح، القلب) وتوافقها وتوازنها.

<table>
<tr><td>والحياة</td><td>العلمية</td><td>المعرفة</td></tr>
</table>

هناك نظريات عدة تفسر نشوء الحياة بعضها يتمثل بسقوط جزئيات عضوية على الارض من المذنبات ويزرع الارض بالحياة من كائنات ذكية في كواكب متقدمة وهناك اتجاهان واضحان في دراسة سر خلق الكون والحياة، اولهما يعتقد بأن هناك غاية في ذلك وثانيهما ينكر وجود الغاية بل الى وجود الصدفة , ومن المواضيع التي تثير الاهتمام بسر خلق الكون والحياة، اصل الكون والتطور والنظريات المختلفة التي تتحدث عن نشوءه والتي يتقبلها كل من العلم والعقل والعلماء والفلاسفة، وكذلك نشأة الحياة والنظريات القديمة والحديثة، المادية منها والمثالية والغربية والشرقية التي تتعامل معها.

تتفق العديد من النظريات المطروحة عن الخلق على ان السموات والارض تكونت من مادة قديمة كانت مبعثرة في الفضاء وان الارض لم تكن موجودة من قبل ومن هذه النظريات:

• نظرية لابلاس (النظرية السديمية) : كانت الكواكب السيارة تدور حول الشمس بشكل سديم منتشر في ارجاء السماء بشكل كتل غازية ملتهبة تدور حول نفسها. تفقد حرارتها بالتدريج ويتقلص حجمها وتزداد سرعتها وتؤدي الى انفصال بعض الاجزاء من وسطها.

- نظرية شامبرلن ومولتون (نظرية الجسيمات الكوكبية): تفترض وجود جسم غريب لم يصطدم بالشمس يؤدي الى حدوث سلسلة من الانفجارات الكبيرة في جسم الشمس.

- نظرية المد الغازي: تفترض هذه النظرية اقتراب النجم من الشمس الى انبثاق لسان هائل من الغاز المشتعل نحو الجهة المقابلة للنجم، تكونت فيما بعد الكواكب السيارة.

- نظرية يفون: وتعبر عن نشأة كواكب المنظومة الشمسية نفسها حينما اصطدمت بكوكب خارجي صغير ادت الى تطاير اجزاء من الشمس.

- نظرية سحابة الغبار العظمي: كانت الشمس وجميع الكواكب في هيئة سحابة هائلة من الغاز نتيجة التكاثف وقوة الجاذبية.

من النظريات المتداولة والاكثر تخصصاً في تكوين الحياة:

- سقوط بعض الجزيئات العضوية على الارض من المذنبات.

- التخصب العام (زرع الارض بالحياة من كائنات ذكية تقطن كواكب متقدمة).

- ظهور الحياة بعد دهور متعاقبة من الترشيح الكيميائي في البحار.

- ظهور الحياة بسرعة وبعد تكون الارض بزمن وجيز.

- ظهور الحياة بتصاعد قطرات ماء مغبرة الى السطح جمعت منها مواد كيميائية تحولت بسرعة الى حياة.

ووفق ذلك يمكن تقسيم هذه النظريات المذكورة وغيرها الى:

* النظريات الحياتية :

تتضمن فكرة النشوء الذاتي للحياة، وقد برزت هذه النظرية عندما اعتقد المصريون القدامى بان الضفادع نشأت من طمى النيل وان قدامى اليونانيين طرحوا فكرة ان الجرذان نشأت من الازبال وان طفيلي الامعاء يتجدد ذاتياً. وايد لويس باستور في (1872) خطأ فكرة النشوء الذاتي وقال (ان الكائنات الحية لا يمكن ان تنشأ الا من كائنات حية مثلها). حيث اثبت هذا العالم بتجاربه العلمية ان الاحياء المجهرية التي تعيش في الماء، كائنات حية مستقلة،

ترد الى الماء من الخارج وتتوالد فيه وتقوم بعملية التخمر. وهكذا ثبت في نهاية المطاف ان الحياة لا تنشأ الا من الحياة وان النطفة بكافة اشكالها لا التوالد الذاتي هي القانون العام السائد في دنيا الاحياء. وان الحياة لا تتوالد ذاتياً بل من وجود قوة خالقة لمادة البروتوبلازمية حيث بثت فيها الحياة.

* النظريات المادية

يؤكد الماديون ان الحياة تولدت من اتحاد مركبات لا عضوية وعضوية بسيطة سادت جو الارض في فترة من الزمن، بعدها تكتلت هذه المواد العضوية لتكوين جزئيات عضوية عملاقة دخلت في سلسلة من التفاعلات المعقدة وكونت البروتوبلازم من هؤلاء الماديين الاكثر حماسة العالم بخنر الذي يؤكد ان الحياة محصورة في البروتوبلازم وان حركتها طبيعية لها مواصفات كيمائية محددة بدرجة حرارة معينة، ولكنها تعدت درجة حرارة هذه فتسمى عندئذ بالحياة الكامنة.

* النظريات الكيميائية

بدأت الارض وكانت خالية من الاوكسجين تتألف من بخار الماء وغاز الميثان وثاني اوكسيد الكاربون وغاز الامونيا، وتم صنع جزئيات ما قبل الحياة من هذه الغازات مختبراً، حيث يتم خلط هواء بدائي في بودقة زجاجية وقصفها بشرارة منتجة احماض امينية ومواد سكرية وقواعد نتروجينية بوساطة صواعق وتمتص اطيان الصلصال الغنية بالنيكل عشرون حامضاً امينياً فقط لتكوين البروتين، ويمكن وضع الاحماض الامينية في احوال حارة وجافة جداً لتكوين قطيرات يطلق عليها طلائع الخلايا ويشير القائمون بالتجربة على ان اصل الحياة سبق اصل الوراثة. انه يمكن خلق الوحدات البنائية والجزئيات الصغيرة في المختبر تحت ظروف بيئية قد تكون مماثلة بالظروف التي كانت سائدة على سطح الارض (الخلق الكيميائي قبل الحياتي). تتم عملية تحول الغازات الى مركبات عضوية (الاحماض الامينية البروتينات)

باستخدام طاقة اشعة الشمس ومنها اشعة فوق البنفسجية والتي تمتص بسهولة من قبل غاز الميثان وبخار الماء والامونيا. كما يمكن استعمال التفريغ الكهربائي، كما اقترح ذلك الباحث يوري، فضلاً عن ذلك قام ميلر بتنظيم تجربة حول التفريغ الكهربائي.

ويعتقد ان ملايين من السنين قد مرت قبل ان تتميز المواد العضوية المعقدة مثل البروتينات والاحماض النووية في ماء المحيطات، حيث تجمعت هذه الجزيئات الكبيرة لتكون كتلاً وانظمة تعمل بها قوى فيزيائية كيمائية ادت الى تكوين تراكيب جديدة اكثر تعقيداً وذات فراغات واعضاء واخيراً تكونت الكائنات الحية البسيطة، ويرى اوبرابن انه نتيجة لتكرار التفاعلات نشأت القدرة على التكاثر الذاتي.

* نظرية ميلر

تتضمن نظرية ميلر الحامض الاميني من المكونات الموجودة في الغلاف الجوي الذي كان يحيط بالكرة الارضية لحظة ولادتها ومنها الامونيا والهيدروجين والميثان وبخار الماء. وقد اجربت التجربة في وعاء خاص (نظام مغلق من الزجاج واقطاب مصنوعة من التنكستين) يوضع فيه المكونات المذكورة حيث يمتزج بخار الماء مع الغازات المختلفة (ميثان، امونيا، هيدروجين) ويعرض الى تفريغ كهربائي، حيث تمر الغازات من خلال منطقة تبريد لاسالة الغازات. وبعد اسبوع من اجراء التفاعل يلاحظ وجود مواد عضوية مثل الاحماض الامينية ومركبات اخرى. تتم عملية التحول (المواد غير العضوية الى مركبات عضوية) بمرحلتين الاولى منها يتكون سيانيد الهيدروجين والادهايدات وفي الثانية تتكون الاحماض الامينية بطريقة ستربكر. اعتمد الباحث على تفاعل ستربكر، حيث حضر الحامض الاميني من تفاعل السيانيد ومركب كربوني في محيط من الامونيا ويتكون في الخطوة الاولى النتريل الاميني الذي يتحلل مائياً في محيط حامض قوي ومن ذلك ينتج حامض الاميني.

وقد اقترح الباحث استعمال بيروكسيد الهيدروجين H_2O2 كعامل مساعد على اعتبار انه ابسط جزيئة بعد الماء، واحتمال وجوده في الكرة الارضية او غاز كبريتيد الهيدروجين H_2S الصادرة عن البراكين الموجودة في ذلك الوقت.

تعرضت نظرية ميلر الى شكوك وخاصة في :

- لم توضح كيفية تحول النتريل الاميني الى حامض اميني في ظروف الارض التي تحتوي على تركيز عال من الامونيا (يتميز جو الارض في بداية تكوينها بكونه جواً مختزلاً تحول الى جو مؤكسد بآلية غير معروفة).

- لم يوضح ميلر مصدر تكوين الاوكسجين في الجو حيث اعتبر حلقة مفقودة.

- ان تطور الارض وتحوله من المحيط ما قبل الحيوي الى محيطه الحيوي قديم وفق التصورات الاتية :

- كانت الارض في بداية تكوينها خالية من طبقة الاوزون وان جزيئات الماء قد تعرضت للاشعة فوق البنفسجية والاشعة الكونية مكوناً بيروكسيد الهيدروجين. بعدها دخلت جزيئة بيروكسيد الهايدروجين في تراكيب الغلاف الجوي المحيط بمعظم الكواكب السيارة الدائرة حول الشمس ذات الظروف البيئية المشابهة لبيئة الارض عند تكوينها. ودخلت في تفاعل تحلل النتريلات الامينية الى حوامضها الامينية طارحة في الجو غاز الاوكسجين والذي يعتبر ناتج عرضي لهذا التفاعل، وتجمعت جزيئات الاوكسجين بعد مرور الاف السنين في بيئة الارض الجوية وتجمع الحامض الاميني في محيطه المائي (البحر) وتعرض للاوكسجين لدى وصوله الى طبقات الارض مكوناً الاوزون الذي يكون طبقة واقية للارض، وتحولت الارض من المحيط المختزل الذي يحتوي على الميثان والهيدروجين والامونيا الى جو مؤكسد يحتوي على الاوكسجين وطبقة اوزون تحميه مع تراكم عال للاحماض الامينية في البحر ولذلك تم خلق جو مناسب للحياة.

نظرية اوبارين

يقول اوبارين ان جميع المحاولات التي اجريت لتوليد الحياة من المواد غير العضوية سواء تحت ظروف طبيعية او مختبرية قد باءت بالفشل. وان الظروف الطبيعية والكيميائية التي سادت على الارض في الطبيعة قبل ظهور الحياة والتي تمت فيها التفاعلات الكيميائية المعقدة والتي ادت الى ظهور تلك الحياة تختلف عن الظروف السائدة الان. كما يشير اوبارين الى ان هذه المواد العضوية بدأت تتجمع في ماء المحيط مؤدياً الى نشأة القدرة على التكاثر الذاتي ولكنه لم يستطع اثبات ذلك. ان تحول المواد غير الحياتية الى حياتية روحية لا تتفق مع المفاهيم الرياضية او قوانين الترموديناميك الحراري وكذلك اليات التفاعلات، فالتحول من مواد لا عضوية الى عضوية ممكن ولكن تكوين مفهوم حي غير ممكن. وقد ذكر اوبارين نفسه ان عملية خلق المادة الحية من مادة عديمة الحياة اضحت الان في عداد المستحيل وذلك لاختلاف الظروف السائدة اليوم على الظروف التي سادت في فجر الحياة كما ذكرنا سابقاً .

الحياة الفكرية للبشر

ان التقدم العلمي الذي احرزه الانسان هو في الواقع نتيجة نموه الفكري في الوجود والكشف عن القوانين المنظمة لهذا الكون بعد ان يتلاحم في ذلك الفكر النظري بالفكر التجريبي. وفي عصرنا الحديث نجد ان الانقسام التقليدي القديم بين الفلسفة العقلية التي تفترض ان العقل هو مصدر المعرفة وبين الفلسفة التجريبية التي تعتبر التجربة هي مصدر المعرفة لازال كبيراً ولم يتلاشى. ومع هذا فنعتقد ان الافكار الاتية قد تساهم في تحديد الحياة الفكرية للبشر:

- ان العقل مقياس كبير في التفكير البشري.
- ان التجربة مهمة كاداة لتطبيق المقياس العقلي، ولكن ليس بمفردها.
- ان الاتجاه العقلي يفترض ان لا يتجاهل دور التجربة في العلوم والمعارف البشرية.
- لا يمكن ان تكون التجربة بذاتها المصدر الاساسي والمقياس الاول للمعرفة

التخصصات العلمية

يمكن تقسيم العلوم الى اقسام مختلفة ومنها الاساسية (الصرفة) والتطبيقية والعلوم الانسانية والعلوم الاساسية تتضمن العلوم الطبيعية باستثناء الهندسة والعلوم التطبيقية، وتشمل الرياضيات والفيزياء، والكيمياء، وعلوم الحياة، وعلوم الارض والفلك والارصاد الجوية.

والعلوم الانسانية تشمل الفلسفة والعلوم الاجتماعية وغيرها اما التخصصات الجامعية فتطورت مع تطور العلوم المختلفة فالتخصصات التقليدية كالهندسة والزراعة والعلوم الصرفة والاجتماعية بدأت في القرن التاسع عشر، وفي القرن العشرين اضيفت تخصصات اخرى مثل ادارة الاعمال والصحافة والاعلام وعلوم المكتبات والاقتصاد والسياسة والشؤون العالمية، وشهد العالم في القرن العشرين ايضاً تطوراً مذهلاً في كل الميادين والاتجاهات العلمية، حتى لم تعد هناك حدود بين مختلف الاختصاصات، فالعلوم الطبية مثلاً تحتاج العلوم الهندسية في اختباراتها الحديثة والاخيرة تعتمد على العلوم الفيزيائية والكيميائية في الاستنباط والتحليل وتعتمد ايضاً على الرياضيات لوضع الاساس الرياضي للبحث او الاختصاص.

ان التقدم العلمي في العلوم كالكيمياء وعلوم الحياة والرياضيات وعلوم الحاسبات والعلوم الاخرى احدثت تطورات ونوعية هائلة في الكيمياء وعلوم الحياة كالتركيب الذري والجزيئي واستخدام اشعة ليزر والاشعة السينية الواطئة الطاقة والعالية منها واشعة كاما وانبثق مثلاً الكيمياء الحيوية وعلم الاحياء الجزيئي والاحيائيات الكيميائية، كما حصلت تطورات هائلة في علم الانزيمات، واصبح لعلم الاحياء عشرات العلوم الفرعية وتطبيقات كثيرة منها الهندسية الوراثية في المجالات الطبية والزراعية والصناعية وانتاج الهرمونات. وكان عام 1997 يمثل حدثاً علمياً مثيراً بولادة النعجة دوللي بطريقة الخلية الجسمية المحقنة الى بويضة غير مخصبة لشاة اخرى بعد ازالة نواتها وزرعها في رحم شاة ثالثة.

ادى التقدم والتطور في العلوم الاساسية والتي تسمى الصرفة على سبيل المثال الى استحداث موضوعات وتخصصات جديدة وعلوم بينية جديدة لم تكن معروفة خلال النصف الاول من القرن الماضي، ومن نتائج ذلك احداث تغييرات واسعة في بنى المناهج الدراسية

والبحثية، كما تحولت هذه التطورات في الجامعات العالمية المتقدمة الى مناهج دراسية وبحثية وشهادات تم تنفيذها بأساليب مختلفة ومنها:

- شطر التخصصات التقليدية.

تم شطر العديد من اختصاصات العلوم الصرفة فمثلا في الكيمياء تكونت اختصاصات جديدة مثل الكيمياء الصناعية والكيمياء الطبية الحياتية وغيرها.

- استحداث الاختصاصات البينية

تم استحداث هذه الاختصاصات في الجامعات العالمية في الفيزياء والكيمياء والرياضيات لاعداد خربجين لبعض القطاعات مثل الكيمياء الهندسية والفيزياء الهندسية والكيمياء الزراعية والطبية.

- الموضوعات المساعدة

اضيفت بعض الموضوعات المساعدة للعديد من اختصاصات العلوم الصرفة ومنها التربية والادبيات وخدمات المكتبة الحديثة واستخدام الاجهزة والانسان الالي والحاسوب.

- الاختصاصات الشطائرية

تم ذلك في الجامعات البريطانية من خلال توسيع سني الدراسة الجامعية الاولية للاستفادة منها في زيادة فرص التدريس المنهجي والتطبيقي وتأهيل الدارسين للعمل في قطاعات الانتاج المختلفة.

ادخلت بعض الجامعات العالمية المناهج الجامعية الحديثة على مستوى الدراسات الاولية كما في بريطانيا والمانيا مثلا وحدثت تطورات مذهلة في مجال العلوم الكيميائية، وخاصة خلال النصف الثاني من القرن العشرين، بما في ذلك التعامل مع التطورات والمحتوى والمفردات والاليات المعروفة وقد ادت هذه التطورات ايضاً لفتح قنوات جديدة في مجال البحث العلمي والصناعات الكيميائية. وتوسعت بعض الجامعات الامريكية بشكل كبير في منح درجة البكالوريوس، وفي الدراسات العليا، في تخصصات الكيمياء المتعلقة بالدواء والزراعة، والهندسة، والعلوم الاحيائية والفيزياء والتعليم، وغيرها كما زاد الاهتمام بالبحث العلمي على

مستوى الدراسة الاولية في بعض الجامعات البريطانية واعداد الاطروحات الجامعية. وادخل الحاسوب في الجامعات ذات الشهرة العالمية ليس فقط كمواضيع للدراسة المتخصصة، بل كوسيلة لتوضيح الهياكل والبنى الكيميائية وتشخيص بعض المركبات .

الرياضيات :

للرياضيات تاريخ طويل امتد الى اكثر من 4000 سنة، اكتشف منها الاعداد غير النسبية والرموز الرياضية والتفاضل والتكامل والهندسة التحليلية. وقد ادت هذه الاكتشافات الى ثورت نوعية منها الهندسة الاقليدية في الثلث الاول من القرن الماضي وظهور نظرية المجموعات غير المنتهية، وظهور مفاهيم رياضية جديدة فاكتشاف الهندسة اللاقليدية ادى الى معرفة التعددية في الرياضيات (هندسات مختلفة) وساهمت نظرية المجموعات الى انظمة مختلفة عددية وجبرية.

والجدير بالذكر ان تطور الرياضيات كان ملازماً للفيزياء، بعدها انفصل الرياضيات وتحرر وبدأ باستحداث انظمة رياضية مجردة مستخدمة نظرية المجموعات والمنطق الرياضي، وبذلك استخدم العاملون والباحثون في الرياضيات، وسائل لايجاد فهم اوسع في الاقتصاد والتغير السكاني والزلازل وسلوك الانسان والطب، استحدث نتيجة هذا التأثير اختصاصات جديدة منها الاقتصاد الرياضي، وعلم الاحياء الرياضي وعلم الارض الرياضي وعلم النفس الرياضي وعلم اللغات الرياضي واعتماد ما يسمى بالنمذجة الرياضية استحدث بعد ذلك اختصاص "الرياضيات التطبيقية" الذي شمل ايضاً رياضيات الفضاء، رياضيات الطاقة، اما الرياضيات الصرفة مثل الجبر الابدالي ونظرية الرمز ونظرية الحلقات ونظرية الاعداد والتوبولوجيا الجبرية والتوبولوجيا التفاضلية والهندسة الجبرية والتفاضل والتكامل والمعادلات التفاضلية والتكاملية فتدخل كارضية لبناء النماذج. اما في الجامعات الامريكية والبريطانية فاحتوت من بداية القرن العشرين والى سنوات سابقة في نهاية القرن العشرين

اقساماً للرياضيات التطبيقية واخرى للرياضيات الصرفة واصبحت في الوقت الحاضر للعلوم الرياضية.

تنمو الرياضيات بسرع متباينة وتتوسع استخداماتها واتجاهاتها في نواحي المعرفة، ومنها بناء النماذج الرياضية، والتعاون مع علماء الحياة واعضاء الكائن الحي وفهم سلوك الانسان والطب لبناء نماذج رياضية تساعد الانسان في السيطرة على بعض الظواهر الطبيعية، وكان من نتيجة هذا ان اصبح بالامكان صياغة الكثير من المسائل الحياتية بواسطة نماذج رياضية دقيقة، تصف عمل القلب والدماغ واعضاء اخرى من الجسم، وتوجد نماذج للآذان والعين، وفي مجال جهاز الدوران طورت طرق حسابية متعددة فضلاً عن ذلك اعتمدت نماذج رياضية لدراسة تأثيرات الهرمونات و ـــــ السائل في الاذن والعصب السمعي والتكهن بالزلازل والتركيب البنائي للـ DNA. ويمكن التطرق الى نظرية الكوارث الرياضية (Catastroph Theory) التي يمكن استخدامها في وصف بعض الظواهر غير المستمرة وهي لها ملامح متعددة منها الجانب الرياضي والجانب التطبيقي والجانب الفلسفي، يمكن اعتمادها في علوم الحياة والفيزياء: اما نظرية الفوضى (Theory of Chaos) فتعني بدراسة رياضية لسلوك الجسم الفوضوي ووصف حركاته وايجاد نظام فوضوي يطبق في الدوائر الكهربائية، الطقس، مسارات الكواكب والاقمار والتفاعلات الكيميائية ونمو الحشرات والاسماك والامراض وانتقال فايروس (HIV) في الايدز وضربات القلب والموجات العصبية.

الحاسوب :

ساهمت الرياضيات في تطور تقانات الحاسوب وتصنيع الحاسوب ذات قدرات حسابية عالية جداً، استحدثت فيها مواضيع متخصصة كاسس رياضية في علوم الحاسوب، رياضيات الانسان الالي، رياضيات الذكاء الصناعي، والهندسة الحاسوبية، ان تأثير الحاسبات لم يقتصر على الرياضيات (العلم الحسابي) وظهور الرياضيات الحسابية وهي الرياضيات التي

تعني بدراسة الخواص المشتركة للنماذج الرياضية وتطوير خوارزميات، بل امتد الى علوم اخرى.

بدأ تطور اجهزة الحاسوب باعتماد البرامج التشغيلية "برمجيات" وهذا التطور المتسارع ادى الى تطور علوم الحاسوب في عدد كبير من الجامعات والكليات لتشكيل حقول منفصلة لعلوم الحاسوب كالمعلومات والتعليم الحاسوبي وعلم التحكم الالي ومعالجة المعلومات ومعالجة البيانات. تتداخل الحاسوبيات مع مواد كثيرة من المواد العلمية والانسانية والعلوم الاساسية والتطبيقية وقد ادى هذا الى التزاوج مع حقول المعرفة المختلفة وخاصة في مجال تطوير التطبيقات الميدانية، وعلى سبيل المثال "اللسانيات الحاسوبية" نتيجة لتداخل علوم الحاسوب مع علم اللغة. ان التخصصات الاكثر اهمية في المستقبل القريب تقانة الحاسوب والروبوتات، على مستوى البرامج ونظم المعلومات وهندسة النظم وهندسة البرامجيات وتقانة المعلومات، ونظم المعلومات والشبكات العصبية والواقع الافتراضي وشبكات الالياف البصرية وأمن المعلومات والتشفير ومعالجة البيانات ونظم التعليم والتعلم ونظم التصميم والانتاج والشبكات العصبية ونظم المعرفة. اما الجامعات العراقية فاستحدثت تخصصات جديدة في مجال الحاسوب، مثل تقانة الحاسوب والروبوتات، وانظمة الهندسة، والذكاء الاصطناعي وادارة المعلومات وامن الحاسوب، ونظم المعلومات، والواقع الافتراضي، ونظم دعم القرار، والترجمة الالية والشبكات العصبية، وحماية شبكات المعلومات والمعرفة الحوسبية ونظم المعلومات، ونظم المعلومات وأمن الحواسيب والمعلومات، يضاف الى ذلك استحدثت كليات متخصصة لاستخدام الحاسوب وانشاء مراكز البحوث ومراكز متخصصة في اللغة العربية والحاسوب وترجمة اللغات.

اختصاصات علوم الحياة

- علم المناعة.

- علم السموم.

- العلوم البيولوجية العصبية (علم الاعصاب).

- الطب الحيوي (العلوم الطبية الحيوية).

- علم الوراثة البشرية.

- علم الاحياء.

- البيولوجيا الجزيئية.

- الكيمياء الحيوية.

- اخلاقيات علم الاحياء (Bioemics).

- مراكز البحوث.

اما مراكز البحوث الحديثة في ضوء البحوث المستقبلية لعلوم الحياة فتتضمن :

- اعادة انشاء مراكز بحوث النخيل.

- مركز بحوث الاصول الوراثية والبذور العراق.

- انشاء مركز للتطبيقات الحاسوب في علوم الحياة.

- انشاء محميات طبيعية جديدة.

- انشاء مركز وطني لبحوث البيئة والتنوع البيولوجي وحماية الاصول الوراثية الوطنية.

- مركز دراسات البيئة المائية ومصايد الاسماك.

- مركز البحوث والدراسات لتنظيف البيئة من التلوث.

- مركز الدراسات والبحوث في مجال الامراض الموروثة في العراق.

علوم الارض :

يتطلب تطوير دراسة علوم الارض في المراحل الاولية وذلك بتحديث المناهج الدراسية والمفردات العامة من المواضيع الرئيسة لعلوم الارض واعتماد الكتب الحديثة.

- تقديم مواضيع التدريس الحديثة مثل الجيولوجيا والبيئة، الجيوفيزياء، الهندسة والنمذجة الرياضية، البرامج والتعدين، زلزالية الطبقات، الجيولوجيا الرسوبية، تحليل البحار،

الاثار الجيولوجيا والرباعية، علوم التربة والكيمياء الجيولوجية العضوية، تطبيقات حديثة في الاستشعار عن بعد والتنقيب عن المعادن، والفيزياء.

■ يقترح استحداث اقسام علمية متخصصة في جيولوجيا البترول والجيولوجيا والموارد المعدنية للصخور الصناعية والمياه والجيولوجيا والموارد، والجيولوجيا البيئية، والهندسة، والجيولوجيا على البحر.

■ تطوير الدراسات العليا وذلك بتعزيز المناهج التطبيقية في الدراسات العليا والتركيو على مواضيع تتعلق بالموارد المعدنية والنفط والمياه، بالاضافة الى التخصصات العلمية البحتة.

■ استخدام مراكز البحوث الحديثة مثل: مركز ابحاث المياه، والنفط مركز البحوث بحوث التصحر مركز الجيولوجيا الهندسية مركز البحوث.

المعلومات الحيوية

قد يكفي هنا ان نقول ان (المعلومات الحيوية) هي العملية التي بمقتضاها تكون العلاقات بين التقانات الاحيائية وتقنيات الحاسوب، وذلك لغرض تبادل المعلومات والتجارب وما يتصل بانتقال الافكار والمعلومات لغرض فهم الحياة وموت الكائنات الحية وتمثل ايضاً التكامل بين الرياضيات والاحصاء والحوسبة وعلوم الحياة لغرض تنظيم (المعلوماتية الحيوية) وتحليلها وتفسيرها.

وعلى الرغم من حداثة علم المعلوماتية الحيوية انه علم شديد التعقيد يستمد اصوله ومسائله من عدم علوم اهمها علوم الرياضيات وعلم الحاسوب فضلاً عن كثير من التوضيحات من العلوم الطبية او الناتجة من دراسات الجينومات البشربة (المحتوى الجيني) للكائنات المختلفة، وفضلاً عن ذلك استعمال المعلومات الالكترونية لفهم الشبكة الوراثية واستحداث مجسمات ثلاثية الابعاد للجزئيات المعقدة من الحاسوب.

اسس المعلوماتية الحيوية

اتخذت المعلومات الحيوية، اشكالاً مختلفة واستخدم اساليب وادوات للمعلومات متنوعة تتفق مع فكرة الوحدات الاساسية المتمثلة بالـ د.ن.أ والجين، الا انها في نفس الوقت تختلف باختلاف تسلسل هذه الوحدات نفسها ومدى بساطتها او تعقدها. ان الظروف والملابسات التي سادت حالات حدوث الامراض المختلفة، وما ارتبط بهذه الظروف من مظاهر التغير الكبرى التي تتمثل الى حد كبير في زيادة الاتجاه نحو تحديث المعلومات حول علم الاحياء البشري، ومنها حالات الشذوذ الكروموسومي والامراض ذات العلاقة فمثلاً تتصف ملازمة دون (Down) بكون الخلية المنفردة تحوي نسخة ثالثة من الكروسوم 21، تتحدد بالفحص المجهري، الا ان التقدم الهائل قد امكن متابعة التغيرات في الـ د.ن.أ ومنها الطفرات المسؤولة عن العديد من الامراض الوراثية. والمهم على اية حال ان المعطيات الحديثة تعتمد اعتماداً كبيراً على التقنيات الحديثة التي يراد توصيلها الى الباحثين وتصبح هذه العمليات من اساسيات علم المعلوماتية الحيوية مرتبطة باجهزة متطورة وفعالة، ومن هنا جاءت أهمية وخطورة هذه الوسائل التي تقدم بغير شك فائدة كبيرة من تقدم هذا العلم ومن اساسياته وتطبيقاته في الصحة والوراثة.

ومن العلاقات المشتركة بين علماء الحاسوب وعلماء الاحياء الجزيئي يمكن قيام عالم الحاسوب بتقديم التفسير الخاص بالقواعد والمعجم اللازم لتشفير الـ د.ن.أ وبالتالي تقدم في المستقبل وبشكل مبين ملائمة المعلومات المهمة لتحديد الكائن الحي. فضلاً عن ذلك هناك امكانية لتطوير الحاسوب باستعمال الـ د.ن.أ واستخدام القواعد كرموز حسابية فعند اضافة كمية محددة من الـ د.ن.أ يمكنها خزن جميع المعلومات الموجودة في حواسيب العالم. ان فكرة الحاسوب الحياتي التي ظهرت في عام 1995 امكنت تطويع الـ د.ن.أ لمعالجة المعلومات. واخيراً فان علم الاحياء لم يعد مقتصراً على المختصين به بل دخلت الى علم الحاسوب وعلمائه الذين بدأو يتعلمون هذه العلوم والمشاركة فيها ضمن فرق بحثية من المختصين لتطوير الحاسوب الحياتي، اذ يمكن تصور علم الاحياء يقع في قلب تحول نموذجي رئيس يقوده علماء الحاسوب وان علم الاحياء اصبح علماً معلوماتياً يسمى

بالمعلوماتية الحيوية وان حجم مكونات الحاسوب الحياتي اصغر بمليارات المرات من حجم شرائح السليكون تتميز بسعة خزن هائلة جداً وسرعة معالجة لحل بعض المسائل المعقدة.

ومن وجهة نظر الحاسوب فان الـ د.ن.أ وتركيبها البنائي يمثل نظاماً ذكياً ورصيناً لخزن المعلومات وان علماء الحاسوب قد تعودوا التعامل مع نظام ثنائي رقمي للتعبير عن الحروف الهجائية والاعداد والرموز وشخصوا مباشرة الحرف الهجائي الرباعي في الـ د.ن.أ وذلك لتشفير الرسائل وان كل تتابع ثلاثي في الـ د.ن.أ عبارة عن نظام معلوماتي لصنع البروتينات وغيرها من المركبات بالرغم من عدم القدرة على حسم موضوع الشكل البنائي الثلاثي للبروتينات بالرغم من وجود نماذج رياضية تخدم الغرض.

ونظراً لكون خزن المعلومات في حواسيب متطورة تستثير الحواس المختلفة بشكل درامي، اذ يعيش الباحث فعلاً داخل الحاسوب وهو يرى الحياة البشرية مخزونة وكلها صور من تقانة العصر الحديث ويعني بالضرورة استغلالها وتطويرها اذ قام العديد من الباحثين في الدول التي ساهمت في نجاح مشروع الجينوم البشري بتطوير هذه الوسائل والقيام بمقارنات اذ تم دراسة التكوين الوراثي في الكائنات الحية غير البشرية كبكتريا القولون وذبابة الفاكهة.

يعد مشروع الجينوم البشري Human Genome Project (HGP) الاداة التطبيقية الرئيسية للمعلومات للعلماء والباحثين والمعلومات الحيوية وذلك على اساس ان الجينات ليست الا رموزاً تدل على اشياء معينة وان استعمالها في المعلومات الحيوية امر حاسم في نجاح البرامج البحثية لغرض توثيق الهوية الاساسية للحياة البشرية على وفق المعلومات المنبثقة عن ذلك تمكنت الكثير من الشركات الصيدلانية او التقانة الاحيائية من تصميم ادوية جديدة لتحسين الصحة البشرية.

ان مبادرة الجينوم البشري جاءت اول مرة في عام 1988 وهدفت الى ايجاد مواقع نحو (100.000) جين بشري في الـ د.ن.أ وان الجينوم البشري يعبر عن 24 زوجاً من الكروموسومات، يتحول الى محتوى معلوماتي عند متابعة تسلسل قواعده تحتاج الى حلول تستند الى علوم الحاسوب وعلوم الرياضيات والاحصاء والعلوم التجريبية. علماً بأن علم

الحاسوب يقدم في الغالب اسهامات بشكل برامج وحلول تتصف بمهارات ادت الى الولوج في اختراع اللغات البرمجية وصف المعلومات التي تنفذ بترتيب معين ويقدم طرائق لوصف العمليات الحيوية المعقدة بعدد من الاسطر البرمجية بدلاً من وصفها بمئات الصفحات باللغة الطبيعية.

وبعد فقد ذكر الباحثون والمراقبون ان القرن الحادي والعشرين سيكون قرن علم الاحياء وان القوة التحليلية الناتجة من (HGP) سيفسر بصورة جذرية جميع البحوث الحيتية والطبية اذ تم:

- البحث عن طبيعة الجينومات وطبيعة تكوينها وتنظيمها في مؤسسات علمية مختلفة.

- التسارع في تنفيذ المشروع الذي خطط له لكي ينجز خلال (15) سنة الا ان التقدم التقني اختصر الوقت الى عشر سنوات.

- البحث في محتوى الجينوم لتعرف نوع المعلومات او مادة الاتصال التي تتضمنها وبالادق تشخيص جميع الجينات البالغة عددها (100.000) في الــ د.ن.أ البشري وكذلك تحديد تسلسلات نحو ثلاثة بلايين من القواعد الكيميائية.

- دراسة طبيعة خزن المعلومات في الحاسوب بتطور متقن وتقنيات تسلسل كفوءة وتطور في الادوات التي تساهم في تحليل المعلومات.

- دراسة التأثيررات التي يراد احداثها في المجتمع والى أي مدى امكن تحقيق ذلك، ونوع الاستجابة.

وعلى الرغم من كل ما قيل عن الدراسات والبحوث التي اجريت من اساليب وتقنيات في المعلومات الحيوية المختلفة وكذلك كثرة ما كتب وما نشر في هذا المجال، لانزال هناك ميادين اخرى كثيرة ومتنوعة للدراسة والبحث. ان بعض هذه الميادين لم يمس حتى الان في القطر خاصة مشاريع الجينوم البشري وهي ميادين ومجالات تجذب اهتمام الباحثين الا انها قليلة معظمها تعالج في الاغلب مشكلات جزئية او فرعية.

ويمكن ان تنقسم العلوم الى قسمين علوم اساسية وعلوم تطبيقية، فالعلوم الاساسية عند البعض جميع العلوم الطبيعية ما عدا الهندسة والعلوم التطبيقية بينما يعتقد البعض الاخر ان العلوم الاساسية المسماة بالصرفة تشمل الرياضيات والفيزياء والكيمياء وعلوم الحياة وعلوم الارض والفلك والارصاد الجوية و يتميز العصر الذي نعيش فيه بتفجر العلوم الصرفة بسرعة مذهلة، وما ان تبتدع عملية جديدة او تعرض فكرة جديدة، حتى يهرع التقانيون الى استغلالها بابتكار جديد فالحاسوب مثلاً كان قبل سنوات موضوع يتكلم عنه الجامعيون صار اليوم الشغل الشاغل في الدوائر والمؤسسات والمصانع. ان التكيف للتفجر العلمي الصرف يفترض تغيير جذري في مفاهيم التعلم والتعليم والمناهج التي تقتضيها الانتقال الى المرحلة الحضارية المتقدمة التي ننشدها لا يمكن ان تنهض بها الجامعة فقط بل المؤسسات ذات العلاقة و ان تفجر العلوم الصرفة يفرض على الجامعات ان تقوم بالقسط الاوفى بحيث تلائم الحاجات المحلية فعلى الجامعات بخاصة ان تضع برامج تعليمية ومناهج تستفاد من البرامج المستوردة، لاننا في عالم سريع التطور يكاد يشمل كل يوم تطوراً علمياً وتقنياً جديداً، ينبغي مثلاً ان يتوافر في الجامعات والمؤسسات وسائل فعالة للحصول على المعلومات ونشرها وان تحديات العصر تفرض على هذه الجامعات ان يكون لها كوادر خاصة

الاتجاهات الحديثة العلمية في العراق

حافات العلوم

تطورت التخصصات الجامعية مع تطور العلوم المختلفة واستحداث تخصصات جديدة، فالتخصصات التقليدية كالهندسة والزراعة والعلوم الصرفة والاجتماعية بدأت في القرن التاسع عشر، وفي القرن العشرين اضيفت تخصصات اخرى مثل ادارة الاعمال والصحافة والاعلام وعلوم المكتبات والاقتصاد والسياسة والشؤون العالمية. ولكل دولة اسلوبها الخاص في تحديد

التخصصات الجامعية وتحديد اعداد الطلاب ونوعية الخريج. وشهد العالم في القرن العشرين ايضاً تطوراً في كل الميادين والاتجاهات العلمية، حتى لم تعد هناك حدود بين مختلف الاختصاصات، فالعلوم الطبية مثلاً تحتاج العلوم الهندسية في اختباراتها الحديثة والاخيرة تعتمد على العلوم الفيزيائية والكيميائية في الاستنباط والتحليل وتعتمد ايضاً على الرياضيات لوضع الاساس الرياضي.

ادى التقدم والتطور في العلوم الصرفة على سبيل المثال الى استحداث موضوعات وتخصصات جديدة وتخصصات وعلوم بينية جديدة لم تكن معروفة خلال النصف الاول من القرن الماضي، ومن نتائج ذلك احداث تغييرات واسعة في بنى المناهج الدراسية والبحثية، كما تحولت هذه التطورات في الجامعات العالمية المتقدمة الى مناهج دراسية وبحثية تم تنفيذها باسباب مختلفة ومنها:

1-البكالوريوس المعتمد على الدراسة والرسالة:

استحدثت بعض الجامعات العالمية مناهج جامعية على مستوى الدراسات الاولية تتضمن الدراسة والبحث معاص ومن هذه الجامعات تلك الموجودة في بريطانيا والمانيا.

2-شطر التخصصات القائمة

تم شطر العديد من اختصاصات العلوم الصرفة فمثلاً في الكيمياء يعطى اختصاص الكيمياء في الصناعة بجانب اختصاص الكيمياء الصناعية والكيمياء الطبية الحياتية مع الكيمياء الحياتية الطبية وغيرها من الاختصاصات في مجال فروع العلوم الصرفة.

3-استحداث الاختصاصات البينية

تم استحداث هذه الاختصاصات في الجامعات الامريكية في الفيزياء والكيمياء والرياضيات لاعداد خريجين في بعض القطاعات مثل الكيمياء الهندسية والفيزياء الهندسية والكيمياء الزراعية والطبية.

4-اضافة الموضوعات المساعدة

اضيفت بعض الموضوعات المساعدة للعديد من اختصاصات العلوم الصرفة ومنها التربية Education والادبيات Literature وخدمات المكتبة الحديثة واستخدام الاجهزة والانسان الالي والحاسوب.

5-الاختصاصات الشطائرية

تم ذلك في الجامعات البريطانية من خلال توسيع سني الدراسة الجامعية الاولية للاستفادة منها في زيادة فرص التدريس المنهجي والتطبيقي وتأهيل الدارسين للعمل في قطاعات الانتاج المختلفة.

وانعكست آثار هذه التطورات بنوعيها الضمني والتوسعي الى تطبيق مناهج الموضوعات الشطايرية في بعض الجامعات البريطانية من اجل زيادة تأهيل خريجي اقسام الكيمياء للعمل في الحقول العلمية والصناعية المختلفة. واخذت بعض الجامعات الامريكية بالتوسع بشكل ملحوظ في منح شهادات البكالوريوس، بل كذلك في الدراسات العليا، في تخصصات مزدوجة تناولت الكيمياء مع الطب او الزراعة او الهندسة او العلوم الاحيائية او الفيزياء او التربية وغيرها. كما زاد الاهتمام بالبحث العلمي على مستوى الدراسة الاولية في الجامعات حتى اصبحت الجامعات البريطانية تطالب الدراسة الاولية في الكيمياء والعلوم الاخرى باعداد رسائل علمية بجانب اجتياز المقررات الدراسية. والتوسع في التطور في العلوم الكيميائية ادى الى شطر بعض الاختصاصات الكيميائية المعروفة الى اختصاصين او اكثر، وهذا من شأنه ادى الى استحداث اختصاصات جديدة لم تكن معروفة قبلاً .

تم التأكيد على وحدة الرياضيات كفلسفة، ويضم هذا القسم اتجاهات متعددة (محاور) مثل رياضيات تطبيقية (هندسية)، علوم حياة رياضية، اقتصاد رياضي، رياضيات مالية احتمالية، احصاء، بحوث عمليات، الرياضيات البحثية وغيرها. كما تم تاكيد تطبيقات

الرياضيات خلال البرامج العامة، وبصورة خاصة اعطاء اهمية خاصة للنمذجة الرياضية، وقد ذهبت بعض الجامعات الى استحداث مشروع (Project) وعيادات رياضية Mathematical Clinics لتاكيد التطبيقات. فضلاً عن ذلك فقد استحدثت دراسات بينية مثل رياضيات واحصاء، رياضيات وفلسفة، رياضيات واقتصاد، رياضيات وفيزياء، رياضيات وهندسة، رياضيات وحاسبات، وغيرها. وادخلت الحاسبات وسيلة لتدريس بعض مواضيع الرياضيات واستنباط طرق تدريس جديدة بديلا عن المحاضرة السائدة حالياً والاهتمام بنشر الثقافة الرياضياتية والوعي الرياضياتي.

تطورت علوم الحاسبات منذ اكتشاف الحاسوب وتصنيعه في مختبرات عدد من الجامعات الامريكية والبريطانية في اوائل الستينات. ولم يكن علم الحاسوب حينذاك قد تبلور علماً مستقلاً بحد ذاته اسوة بالعلوم الصرفة او التطبيقية، ولكن كانت نشأته في احضان اقسام علمية وبالاخص اقسام الرياضيات واقسام الهندسة الكهربائية، حيث ان مواد الحاسوب ومناهجه لم تكن بالوسع والكفاية اللازمة لقيام اقسام مختلفة، كما تركزت مواد الحاسبات باتجاهين احدهما هندسي له علاقة بمكونات ومعمارية الحاسوب والاتجاه الآخر برمجي، حيث يدرس عدداً من لغات البرمجة في كليات العلوم والهندسة، وتستخدم لحل المسائل الرياضية والحسابية والهندسية والاحصائية او لاغراض النمذجة والمحاكاة. وبتطور اجهزة الحاسوب من حيث الاجهزة المادية (Hardware) والبرامجيات التشغيلية (Software) بدأت تتسع المواد العلمية والنظرية العلمية وهذا التوسع المطور والمتسارع ادى الى بلورة علم الحاسبات بوصفه احد العلوم الاساسية، وقام عدد كبير من الجامعات والكليات بتشكيل اقسام مستقلة بعلم الحاسوب Computer Sscience او علم الحاسوب والمعلومات Computer and Information Science او تعليم الحاسوب Computer Education او السيبرنتك Cybernetics او معالجة المعلومات Data Processing او تحت عناوين اخرى تشير الى احد انواع معالجة المعلومات.

توسع تعليم الحاسوب حتى تداخلت مواده مع عدد كبير من المواد العلمية والانسانية واصبحت مواد الحاسوب جزءً من متطلبات مناهج مختلف مواد العلوم الصرفة والتطبيقية والانسانية. وقد نتج عن تزواج علم الحاسوب مع علوم اخرى حقول معرفية وعلمية جديدة وخاصة في مجال التطبيق فمثلاً تطور حقل الالسنيات الحاسوبية Computational Linguistics كنتيجة لتداخل هلم الحاسوب مع علم اللغة او حقل معرفي يدعى نظم دعم القرار Decision Support System كتداخل بين علم الحاسوب او علم نظم المعلومات وعلم الادارة وبحوث العمليات والاحصاء..الخ.

التوجهات الحديثة والمهمة في العلوم

الكيمياء

حدثت تطورات مذهلة في العلوم الكيميائية لاسيما خلال النصف الثاني من القرن الماضي، منها ضمنية واخرى بينية. فالتطورات الضمنية تناولت المضامين والاليات تفسيرات جديدة في الظواهر والاحداث والتفاعلات الكيميائية , نشأت من جراء ذلك ايضاً موضوعات وتخصصات جديدة في ضمن علم الكيمياء نفسه , وادت هذه التطورات كذلك الى فتح قنوات جديدة في البحث لعلمي والتقانة الكيميائية وانشاء صناعات كيميائية جديدة، او صناعات كيميائية بتقنيات جديدة.

اما تطورات النوع الثاني من العلوم الكيميائية البينية فقد تناولت التخصصات التي تربط العلوم الكيميائية بالعلوم الصرفة والتطبيقية المختلفة وادت هذه التطورات الى استحداث اختصاصات او علوم جديدة لم تكن معروفة من قبل واخذت هذه الاختصاصات والعلوم مكانتها اللائقة في البرامج والمناهج والخطط الدراسية والبحثية. ولا يزال بعض هذه التخصصات

او العلوم المستحدثة غير مكتملة الملامح في الوقت الحاضر الا ان التوقعات العلمية تشير الى قرب اكتمالها او رسوخها مع مطلع هذا القرن.

وانعكست اثار هذه التطورات بنوعيها الضمني والبيني الى تطبيق مناهج الموضوعات الشطيرية للعمل في الحقول العلمية والصناعية المختلفة. واخذت بعض الجامعات الامريكية بالتوسع بشكل ملحوظ في منح شهادات بكالوريوس، بل كذلك في الدراسات العليا، في تخصصات مزدوجة تناولت الكيمياء مع الطب او الزراعة او الهندسة او العلوم الاحيائية او الفيزياء او التربية وغيرها.

كما زاد الاهتمام بالبحث العلمي على مستوى الدراسة الاولية في الجامعات باعداد رسائل بجانب اجتياز المقررات الدراسية. والتوسع والتطور اللذين حدثا في المفاهيم في العلوم الكيميائية اديا كذلك الى شطر بعض الاختصاصات الكيميائية المعروفة الى اختصاصين او اكثر، وهذا من شأنه استحداث اختصاصات جديدة لم تكن معروفة قبلاً.

واخذت الجامعات العالمية المعروفة بادخال الحاسوب ليس فقط في موضوعات دراسية تخصصية في الكيمياء، بل ايضاً باتخاذها وسائل ايضاح ضرورية في التدريسات الكيميائية لغرض توضيح وعرض التراكيب والبنى الكيميائية وتماثلاها وترتيبها الفراغية وطرائق تحضير وتشخيص بعض المركبات، وذلك لاغراض انواع المعالجات والحسابات الكيميائية الضرورية في يومنا هذا.

والتوسع في التطور في العلوم الكيميائية ادى الى شطر بعض الاختصاصات الكيميائية المعروفة الى اختصاصين او اكثر، وهذا من شأنه ادى الى استحداث اختصاصات جديدة لم تكن معروفة قبلاً. حدثت تطورات مذهلة في العلوم الكيميائية لاسيما خلال النصف الثاني من القرن الحالي، منها ضمنية واخرى بينية. فالتطورات الضمنية تناولت المضامين والمفردات والاليات المعروفة في الكيمياء وتقديم تفسيرات محسنة او جديدة في الظواهر والاحداث والتفاعلات الكيميائية. وتنشأ من جراء ذلك ايضاً موضوعات وتخصصات جديدة ضمن علم

الكيمياء نفسه. وادت هذه التطورات كذلك الى فتح قنوات جديدة في البحث العلمي والتقانة الكيميائية الى انشاء صناعات كيميائية جديدة، او صناعات كيميائية بتقنيات جديدة.

اما تطورات النوع الثاني من العلوم الكيميائية البينية فقد تناولت التخصصات التي تربط العلوم الكيميائية بالعلوم الصرفة والتطبيقية المختلفة. وادت هذه التطورات الى استحداث اختصاصات او علوم جديدة لم تكن معروفة من قبل. واخذت هذه الاختصاصات والعلوم مكانتها اللائقة في البرامج والمناهج والخطط الدراسية والبحثية في الكثير من جامعات العالم الرصينة. ولا يزال بعض هذه التخصصات او العلوم المستحدثة غير مكتملة الملامح في الوقت الحاضر الا ان التوقعات العلمية تشير الى قرب اكتمالها ورسوخها مع مطلع القرن المقبل. وانعكست اثار هذه التطورات بنوعيها الضمني والبيني الى تطبيق مناهج الموضوعات الشطائربة في بعض الجامعات البريطانية من اجل زيادة تاهيل خريجي اقسام الكيمياء للعمل في الحقول العلمية والصناعية المختلفة. واخذت بعض الجامعات الامريكية بالتوسع بشكل ملحوظ في منح شهادات بكالوريوس، بل وكذلك في الدراسات العليا، في تخصصات مزدوجة تناولت الكيمياء مع الطب او الزراعة او الهندسة او العلوم الاحيائية او الفيزياء او التربية وغيرها. كما زاد الاهتمام بالبحث العلمي على مستوى الدراسة الاولية في الجامعات حتى اصبحت بعض الجامعات البريطانية تطالب طلبة الدراسة الاولية في الكيمياء والعلوم الاخرى باعداد رسائل علمية بجانب اجتياز المقررات الدراسية. والتوسيع والتطور اللذين حدثا في المفاهيم في العلوم الكيميائية اديا كذلك الى شطر بعض الاختصاصات الكيميائية المعروفة الى اختصاصين او اكثر، وهذا من شأنه استحداث اختصاصات جديدة لم تكن معروفة قبلاً. واخذت الجامعات العالمية المعروفة بادخال الحاسوب ليس فقط كموضوعات دراسية تخصصية في الكيمياء، بل وايضاً كوسائل ضرورية في التدريسات الكيميائية لغرض توضيح وعرض التراكيب والبنى الكيميائية وتماثلاتها وترتيباتها الفراغية وطرائق تحضير وتشخيص بعض المركبات، وذلك لاغراض اجراء انواع المعالجات والحسابات الكيميائية الضرورية في يومنا هذا. كما وتم تطعيم مناهج الكيمياء في الجامعات العالمية في موضوعات مساعدة

جديدة لم تكن معروفة من قبل ويعتقد انها اصبحت اليوم من موضوعات ومكملات الثقافة العلمية الضرورية

بعض الاتجاهات الحديثة في الكيمياء والكيمياء الحياتية

بعض الاتجاهات الحديثة في الكيمياء

من الصعوبات التي يواجهها الباحث لاحتواء الاتجاهات الحديثة في العلوم هو تعددها ومحدودية المجال لعرضها. وبرغم ذلك فيمكن استعراض هذه الاتجاهات باختصار شديد ومن ثم التركيز على بعضها وبذلك تعم الفائدة ويرتكز الجهد منعاً لتبعثره. ومن مجالات الكيمياء الحديثة مثلاً ما يأتي:-

- المركبات الحلقية العيانية.

لهذه المركبات القابلية لفصل العناصر الكيميائية واستخلاصها بكفاءة عالية، تستخدم كنماذج لدراسة انتقال الايونات عبر جدار الخلية ولها كفاءة عالية كمضادات حياتية.

- متعددة الجزيئات (البوليمرات).

تحتوي على عناصر ذات القدرة العالية للتوصيل الكهربائي وتنقل الدواء الى المكان المحدد.

- احداث تفاعلات انتقالية، لصنع مركبات ذات تأثير على زيادة الفعالية الحياتية لغرض العلاج.

- استخدام الجزيئات الحياتية من اجل تصنيع الرقاقة الحياتية في حاسبات الجيل القادم.

- تصنيع مواد كيميائية عضوية فائقة التوصيلية.

- تطوير جيل جديد من الحاسبات الفائقة والاجهزة الحساسة، يمكن زراعتها داخل الجسم البشري لاصلاح بعض الانسجة التالفة.

- تطوير اجهزة مطيافية لمتابعة تفاعلات ما قبل الحياة.

- صنع جزيئات كيميائية جديدة للعلاج من الامراض (الايدز والسرطان).

تمهيد للاتجاهات الحديثة في الكيمياء الحياتية

تبحث الكيمياء الحياتية في الصفات الكيميائية والفيزيائية لمكونات الخلية والملامح للانظمة الحياتية لهذه المكونات، وكذلك تفسير ماهية هذه الانظمة في الخلية بصورة دقيقة.

قدمت الكيمياء الحياتية الكثير من الانجازات، فقد ساعدت في توضيح آلية الادوية وساهمت في تشخيص وعلاج الكثير من الامراض وقدمت التقنيات التي امكن استعمالها القياس مستوى الكثير من المركبات الموجودة في الجسم الحي.

يتجاوز عمر الكيمياء الحياتية القرن من الزمن وله تخصصات مختلفة بعضها يتعلق بدراسة المواد التي تتكون منها الخلية النباتية ويسمى عندئذ بالكيمياء الحياتية النباتية والذي يتعلق بالخلية الحيوانية واذا كانت الخلية البشرية هي المقصودة سواء كانت طبيعية ام مرضية فيسمى بالكيمياء السريرية وقد توسعت الكيمياء الحياتية فاصبحت تشمل الكيمياء الحياتية الفيزيائية والكيمياء الحياتية العضوية والكيمياء الحياتية اللاعضوية وكذلك كيمياء التغذية.

يهتم علم الكيمياء الحياتية الحديث بوظائف الانظمة الحياتية، فقد ساهمت وسائل الدراسة في القرن الماضي على ملاحظة هذه الانظمة مباشرة اثناء عملها، اما في الوقت الحاضر فقد تغيرت الصورة واصبح من الممكن الحصول على الملاحظات المفضلة الاكثر تبحراً بواسطة التقنيات القابلة للتطور (المجهر الالكتروني، النظائر المشعة، المناعة، الاطياف).

وقد اعتقد العلماء في نهاية القرن التاسع عشر بانه من الممكن الحصول على بعض المعلومات التي تتعلق بالانظمة الحياتية، وذلك بدراسة كيمياء الخلايا، ولعقود تبعتها اعتمد الكيميائيون الحياتيون على الطرق الكيميائية المتوفرة لديهم والتي افلحت في الحصول على تطورات مفيدة. ان التحسينات المهمة لاساليب التقنية الكيميائية كاستعمال النظائر المشعة زادت بصورة كبيرة من حساسية تشخيص الانواع المختلفة من الجزيئات الحياتية وغيرها وعندما اصبح من الضروري فصل مكونات التفاعل الكيميائي الحياتي وبطرق حساسة جداً، استعملت عندئذ الكروموتوغرافيا والترحيل الكهربائي بصورة تقليدية.

عندما اتجه اهتمام الفيزيائيين والكيميائيين نحو علم الحياة (وربما يعود السبب الى قابلية الخلايا الحية لتكوين نظام، ومع ان القوانين الفيزيائية تؤكد على وجود ميل في الكون نحو عدم الانتظام) برزت عندئذ اساليب التقنية الفيزيائية والكيميائية الفيزيائية، مثل علم الديناميك المائي، علم الاطياف، الحيود لتطبق في مجال علم الاحياء.

ان التقدم الذي حصل في الكيمياء الحياتية قد بدأ بالاعتراف بان الانظمة الحياتية تحتوي على جزيئات صغيرة تهتم الكيمياء العضوية بدراستها وتوضيحها وكذلك على جزيئات ضخمة يطلق عليها بالجزيئات العيانية والتي لا تقل اوزانها الجزيئية عن (100) مليون مرة بقدر كتلة ذرة الهيدروجين. تحتل اهمية الجزيئات العيانية للانظمة الحياتية بقدرتها الخصوصية تجاه التفاعلات الحياتية وفي تكوينها الوحدات البنائية، ويمكن القول وبوضوح بانه قد بذلت في السنين الماضية جهود ضخمة لتوصيف وضم الجزيئات العيانية وكذلك المفاعلات التي تحدث بينها ويحتاج لذلك طرق متقدمة للفصل والتنقية وتوصيف الجزيئات العيانية وذلك للحصول على المعلومات عن التركيب البنائي للجزيئة العيانية.

كان هدف الكيمياء الحياتية لنصف قرن تقريباً هو جمع وتنظيم التفاعلات الكيميائية التي تحدث في الخلايا الحية. وكان الحافز لهذا الجهد الكبير هو ان عدداً مهماً من صفات الخلايا الحياتية يمكن ان يفهم من خلال هذه التفاعلات التي تتميز عادة بتكوين او تكسير الاواصر التساهمية. وقد تم توضيح عملية تحرير الطاقة نتيجة التكسر الكيميائي وكذلك عمليات تحول الجزيئات الحياتية المتبادل وعلميات تجميع الاحماض الامينية والنكيلوتيدات والسكربات والدهنيات لتكوين الجزيئات العيانية.

وخلال الثلاثين سنة الماضية اتضح بصورة جلية ان المفاعلات التي تحدث بين الجزيئات بسبب فيزيائي أي تلك التي لا تكون او تكسر الاواصر التساهمية لها نفس اهمية التفاعلات الكيميائية (أي درجة المسموح لها بالحدوث) تجز بواسطة التغيرات الفيزيائية التي تحدث في بنية (بناء) الجزيئات الكبيرة، وكذلك في تهيئة المراكز النشطة في هذه الجزيئات والناتجة من الترابط غير التساهمي للجزيئات الصغيرة، اضافة الى ذلك فان العديد من الصفات الخاصة

لمجاميع الجزيئات العيانية الموجودة في الخلايا او في الكائن الحي (لغشاء الخلية وجدران الخلايا والكروموسومات).

تعددية الجزيئات الحياتية

تتكون بنية التركيب البنائي الاولي للتراكيب المتعددة الجزيئات من انواع مختلفة من الوحدات المرتبة (المتسلسلة)، مثال ذلك تسلسل الاحماض الامينية الموجودة في البروتينات وتسلسل بواسطة التحاليل الكيميائية. اما التركيب البنائي الثانوي فينطوي على تكوين معقد ذو التراكيب الثلاثية الابعاد ويطلق على توجيه كل وحدة من الوحدات المتعددة الجزيئات بالنسبة الى الوحدات الاخرى ويسمى التركيب البنائي الثانوي كتقليد او عرف بالوضعية (الهيئة والصورة) او وضعية الهيكل الاساس او العمود الفقري المتمثل بالسلاسل البتيدية المتعددة. اما الاشكال التي تتكون فتتمثل بالحلزون والحلقات والسطوح والاعمدة او من انواع مختلطة من هذه الاشكال ويطلق على اتجاه (وضعية) السلاسل الجانبية النسبي (الاحماض الامينية او قواعد الاحماض النووية) بالتركيب البنائي الثلاثي. وتتفاعل (نلتحم) الكثير من المتعددة الجزيئات الحياتية مع بعضها لتكون تراكيب بنائية معقدة مثل عديدة الوحدات كالبروتينات والفيروسات والاغشية والشعيرات.

وتكون الاواصر البتيدية عادة في مستوى واحد، حيث تحدد بانواع التراكيب الثنائية للبروتينات. ومن ناحية اخرى فالاواصر التي تشمل الكاربون الفا تكون تسمح لانواع كثيرة من التراكيب البنائية الممكنة. اما الاواصر الفوسفانية ثنائية الاستر الموجودة في الاحماض النووية فهي قابلة للانثناء ايضاً، وبسبب كون القواعد مرنة ومن حلقات كارهة الماء وفي مستوى واحد محاطة بمجاميع قليلة ذات شحنة لذا فهي تتواجد عادة واحدة فوق الاخرى، مما يقلل من التصاقها بالماء، ويزيد هذا من صلابة التركيب البنائي.

اما متعدد الجزيئات الحياتي الخطي الذي ليس له دوران حر حول الاواصر والتي لا تتفاعل مجاميعه الجانبية فيسمى بالملف العشوائي الذي لا يمتلك بنائيا بابعاد محددة او حجماً

متميزاً، يلتف بواسطة الحركة البروانية. ويمكن قياس الحجم بقيمة تساوي معدل نصف القطر للدوران حول نقطة ما او محور.

$$R_G = \sqrt{\frac{\sum R_1^2}{N}}$$

حيث N عبارة عن عدد المقاطع او الوحدات

R_1 تمثل معدل مساحة المقطع

R_G يتناسب مع $\sqrt{N}$

لكثير من متعددات الجزيئات الحياتية تراكيب بنائية حلزونية وعند تطبيق ذلك على البروتينات نرى ان الاواصر تدور في مستوى واحد حول نقطة معينة ويكون هذا الدوران من مستوى الى مستوى اخر ثابتاً. وفي النيكليونيد المتعدد تتكدس قواعد النيكلونيد المتواجدة في مستوى واحد فوق بعضها البعض وبقابلية دوران قليلة جداً، حيث نجد في الـ د.ن.أ ان خيطي (شريطي) النلكليونايد المتعدد تمتد نتيجة هذا بالتكديس حيث يرتبط كليهما بالاصرة الهيدروجينية لتكوين الحلزون المزدوج. اما بعض الجزيئات المتعددة الجزيئات الاخرى كالبروتين كولاجين فتكون حلزونيا ذا خيوط ثلاثية وتمثل الجزيئات الحلزونية (الفرد والمتعدد الخيوط) امثلة على الجزيئات الممتدة او ذات الشكل الشبيه بالقضيب.

ان الاصطلاح المسمى بالتركيب البنائي الفطري لمتعددات الجزيئات الحياتية كثير الاستعمال، الا انه صعب التعريف، ويمكن ان يقصد به الامور الاتية:

• التركيب البنائي لمتعدد الجزيئات الحياتي كما هو موجود في الطبيعة.

• التركيب البنائي لمتعدد الجزيئات الحياتي بعد استخلاصه على ان يبقى محتفظاً بنشاطه الانزيمي.

• شكل لمتعدد الجزيئات الحياتي الذي لا يمتلك نشاط حياتي الا انه يملك تركيب بنائي ثانوي.

اما الاصطلاح المسمى بالمتغير صفاته الطبيعية (الممسوخة) فهو ايضاً غامض ويقصد به شكل المتعدد الجزيئات الحياتي الذي يملك اقل تركيب ثانوي من الشكل الاصلي (الفطري) ففي البروتينات يقصد بها الملف العشوائي او القريب من العشوائية. اما لهذا المتعدد الجزيئي الحياتي الذي يملك الاواصر الهيدروجينية او الذي لا يحتوي على الاواصر الهيدروجينية الداخلية الذي يمكن ان يجتمع بصورة عشوائية مع اشرطة منفردة مختلفة او شريط منفرد واحد، وعند فقدان بعض من التراكيب البنائية الاصلية (الفطرية) لمتعدد الجزيئات فيمكن ان نعتبره ممسوخاً بصورة جزئية. ويطلق ايضاً على الانتقال من التركيب البنائي المنظم الى التركيب البنائي غير المنظم بانتقال الملف الحلزوني او ان الاصلاح (المسخ) والتغير في صفاته الطبيعية يمكن استعماله بنفس القدرة. ويمكن متابعة وتشخيص التغير في بعض الصفات الفيزيائية لجزيئة مثل اللزوجة الذاتية الامتصاصية ومعامل التركيب، كما يمكن حث عملية الانتقال بواسطة درجة حرارة، الاس الهيدروجيني، المسخ بتركيب معين، اما المواد الكيميائية التي تسبب التحول من الشكل الاصلي (الفطري) الى الشكل الممسوخ للبروتينات مثل اليوريا وكلوريد الكوانيديوم.

وعندما يتكون التركيب البنائي من وحدات ثانوية فتسمى عملية فصل هذه الوحدات بالمسخ وفي بعض الحالات او في كثير من الحالات بالتحليل، ويستعمل الاخير مثلا لتوضيح البروتينات المتكونة من سلاسل ببتيدية متعددة والتي يمكن ان تختزل الى خليط من سلاسل منفصلة.

بعض الاتجاهات الحديثة في الكيمياء والكيمياء الحياتية بعض الاتجاهات الحديثة في الكيمياء

من الصعوبات التي يواجهها الباحث لاحتواء الاتجاهات الحديثة في العلوم هو تعددها ومحدودية المجال لعرضها. وبرغم ذلك فيمكن استعراض هذه الاتجاهات باختصار شديد

ومن ثم التركيز على بعضها وبذلك تعم الفائدة ويرتكز الجهد منعا لتبعثره. ومن مجالات الكيمياء الحديثة مثلا ما يأتي: –

• المركبات الحلقية العيانية.

لهذه المركبات القابلية لفصل العناصر الكيميائية واستخلاصها بكفاءة عالية، تستخدم كنماذج لدراسة انتقال الايونات عبر جدار الخلية ولها كفاءة عالية كمضادات حياتية.

• متعددة الجزيئات (البوليمرات).

تحتوي على عناصر ذات القدرة العالية للتوصيل الكهربائي وتنقل الدواء الى المكان المحدد.

• احداث تفاعلات انتقالية، لصنع مركبات ذات تأثير على زيادة الفعالية الحياتية لغرض العلاج.

• استخدام الجزيئات الحياتية من اجل تصنيع الرقاقة الحياتية في حاسبات الجيل القادم.

• تصنيع مواد كيميائية عضوية فائقة التوصيلية.

• تطوير جيل جديد من الحاسبات الفائقة والاجهزة الحساسة، يمكن زراعتها داخل الجسم البشري لاصلاح بعض الانسجة التالفة.

• تطوير اجهزة مطيافية لمتابعة تفاعلات ما قبل الحياة.

• صنع جزيئات كيميائية جديدة للعلاج من الامراض (الايدز والسرطان). تمهيد للاتجاهات الحديثة في الكيمياء الحياتية

تبحث الكيمياء الحياتية في الصفات الكيميائية والفيزيائية لمكونات الخلية والملامح للانظمة الحياتية لهذه المكونات، وكذلك تفسير ماهية هذه الانظمة في الخلية بصورة دقيقة.

قدمت الكيمياء الحياتية الكثير من الانجازات، فقد ساعدت في توضيح آلية الادوية وساهمت في تشخيص وعلاج الكثير من الامراض وقدمت التقنيات التي امكن استعمالها القياس مستوى الكثير من المركبات الموجودة في الجسم الحي.

يتجاوز عمر الكيمياء الحياتية القرن من الزمن وله تخصصات مختلفة بعضها يتعلق بدراسة المواد التي تتكون منها الخلية النباتية ويسمى عندئذ بالكيمياء الحياتية النباتية والذي

يتعلق بالخلية الحيوانية واذا كانت الخلية البشرية هي المقصودة سواء كانت طبيعية ام مرضية فيسمى بالكيمياء السريرية وقد توسعت الكيمياء الحياتية فاصبحت تشمل الكيمياء الحياتية الفيزيائية والكيمياء الحياتية العضوية والكيمياء الحياتية اللاعضوية وكذلك كيمياء التغذية.

يهتم علم الكيمياء الحياتية الحديث بوظائف الانظمة الحياتية، فقد ساهمت وسائل الدراسة في القرن الماضي على ملاحظة هذه الانظمة مباشرة اثناء عملها، اما في الوقت الحاضر فقد تغيرت الصورة واصبح من الممكن الحصول على الملاحظات المفضلة الاكثر تبحرا بواسطة التقنيات القابلة للتطور (المجهر الالكتروني، النظائر المشعة ، المناعة، الاطياف).

وقد اعتقد العلماء في نهاية القرن التاسع عشر بانه من الممكن الحصول على بعض المعلومات التي تتعلق بالانظمة الحياتية، وذلك بدراسة كيمياء الخلايا، ولعقود تبعتها اعتمد الكيميائيون الحياتيون على الطرق الكيميائية المتوفرة لديهم والتي افلحت في الحصول على تطورات مفيدة. ان التحسينات المهمة لاساليب التقنية الكيميائية كاستعمال النظائر المشعة زادت بصورة كبيرة من حساسية تشخيص الانواع المختلفة من الجزيئات الحياتية وغيرها وعندما اصبح من الضروري فصل مكونات التفاعل الكيميائي الحياتي وبطرق حساسة جدا، استعملت عندئذ الكروموتوغرافيا والترحيل الكهربائي بصورة تقليدية.

عندما اتجه اهتمام الفيزيائيين والكيميائيين نحو علم الحياة (وربما يعود السبب الى قابلية الخلايا الحية لتكوين نظام، ومع ان القوانين الفيزيائية تؤكد على وجود ميل في الكون نحو عدم الانتظام) برزت عندئذ اساليب التقنية الفيزيائية والكيميائية الفيزيائية، مثل علم الديناميك المائي، علم الاطياف، الحيود لتطبق في مجال علم الاحياء.

ان التقدم الذي حصل في الكيمياء الحياتية قد بدأ بالاعتراف بان الانظمة الحياتية تحتوي على جزيئات صغيرة تهتم الكيمياء العضوية بدراستها وتوضيحها وكذلك على جزيئات ضخمة يطلق عليها بالجزيئات العيانية والتي لا تقل اوزانها الجزيئية عن (100) مليون مرة بقدر كتلة ذرة الهيدروجين. تحتل اهمية الجزيئات العيانية للانظمة الحياتية بقدرتها الخصوصية تجاه

التفاعلات الحياتية وفي تكوينها الوحدات البنائية، ويمكن القول وبوضوح بانه قد بذلت في السنين الماضية جهود ضخمة لتوصيف وضم الجزيئات العيانية وكذلك المفاعلات التي تحدث بينها ويحتاج لذلك طرق متقدمة للفصل والتنقية وتوصيف الجزيئات العيانية وذلك للحصول على المعلومات عن التركيب البنائي للجزيئة العيانية.

كان هدف الكيمياء الحياتية لنصف قرن تقريبا هو جمع وتنظيم التفاعلات الكيميائية التي تحدث في الخلايا الحية. وكان الحافز لهذا الجهد الكبير هو ان عددا مهما من صفات الخلايا الحياتية يمكن ان يفهم من خلال هذه التفاعلات التي تتميز عادة بتكوين او تكسير الاواصر التساهمية. وقد تم توضيح عملية تحرير الطاقة نتيجة التكسر الكيميائي وكذلك عمليات تحول الجزيئات الحياتية المتبادل وعلميات تجميع الاحماض الامينية والنكيلوتيدات والسكربات والدهنيات لتكوين الجزيئات العيانية.

وخلال الثلاثين سنة الماضية اتضح بصورة جلية ان المفاعلات التي تحدث بين الجزيئات بسبب فيزيائي أي تلك التي لا تكون او تكسر الاواصر التساهمية لها نفس اهمية التفاعلات الكيميائية (أي درجة المسموح لها بالحدوث) تنجز بواسطة التغيرات الفيزيائية التي تحدث في بنية (بناء) الجزيئات الكبيرة، وكذلك في تهيئة المراكز النشطة في هذه الجزيئات والناتجة من الترابط غير التساهمي للجزيئات الصغيرة، اضافة الى ذلك فان العديد من الصفات الخاصة لمجاميع الجزيئات العيانية الموجودة في الخلايا او في الكائن الحي (لغشاء الخلية وجدران الخلايا والكرو موسومات).

تعددية الجزيئات الحياتية

تتكون بنية التركيب البنائي الاولي للتراكيب المتعددة الجزيئات من انواع مختلفة من الوحدات المرتبة (المتسلسلة)، مثال ذلك تسلسل الاحماض الامينية الموجودة في البروتينات وتسلسل بواسطة التحاليل الكيميائي

التطورات الحديثة في جزيئات الاحماض النووية

تعتبر الاحماض النووية مركبات ذات وحدات جزيئية عديدة تسمى بمقررات الكائنات الحية الوراثية وتتكون من وحدات بنائية تسمى بالنكليوتيدات والنكليوتيدات ناقصة الاوكسجين والتي تتكون من مقاطع ثلاثة، السكر الخماسي بنوعه الرايبوزي ومنقوص الاوكسجين. اما النكليوسيدات فتتكون من السكر والقاعدة النتروجينية.

وقد لعب التركيب البنائي للـ د.ن.أ. الاولي الذي تم تسميته بالحلزون المزدوج ادوارا وظيفية عديدة واساسا لتراكيب بنائية اخرى منها الشكل Z (المتعرج). يضاف الى ذلك فيوجد الـ د.ن.أ. بشكل حلقي في بعض انواع البكتربا والفيروسات وفي بعض انواع اخرى يكون بشكل خيطي ذات نهايات متكسرة ويغلب على جزيئات الـ د.ن.أ الحلقية بلف فائق اللف، ويضيف ان للاخير اهمية حياتية.

النكليوسيدات والنكليوتيدات

تلعب النكليوسيدات والنكليوتيدات ادوارا مميزة في العديد من تفاعلات البناء الحياتي والسيطرة الحياتية في الخلية الحية، حيث تخدم كوحدات بنائية لكل من الـ ر.ن.أ. والـ د.ن.أ بصيغة نكليوتيدات بيورينية وبيربمدينية وكذلك في تكوين مصدر الطاقة بصيغة الادينوسين ثلاثي الفوسفات وجزء من مساعدات الازيم ومنظمات الوستيربة للانزيمات ومراسلات ثانوية الاي. ام. بي الحلقي "cAMP" والجي. ام. بي. الحلقي "cGMP".

تم تحوير العديد من النكليوسيدات الموجودة في الطبيعة كيميائيا واصبحت ذات اهمية تطبيقية وخاصة في المجال الطبي ويتضمن التحوير الكيميائي الجزء الحلقي او السكري مؤديا الى بناء متشابهات تعمل كمضادات للمركبات الايضية ويتضمن Anti-Metabolites و بالتالي يمكن استثمارها في تطور العديد من المركبات التي ربما تتداخل بصورة انتقائية مع وظائف الفيروسات او الخلايا السرطانية، وتقوم المتشابهات النكليوسيدية بدور مهم في العلاج الكيميائي وفي استعمالها في التنميط المناعي او الفيروسات او الخلايا السرطانية، وتقوم المتشابهات النكليوسيدية بدور مهم في العلاج الكيميائي وفي استعمالها في التنميط المناعي او

السيطرة على التعبير الجيني والتي تشكل توجهات حديثة في العلاج. ومنها مثلا استعمال المركب 2,3 Dideoxnucleosides وكذلك (AZT) 3-azido 3-Deoxythymidine وهي من المتشابهات النكليوسيدية التي تستعمل في تثبيط فيروس العوز المناعي البشري التي تسبب Human Immuno HIV Deficiency Virus متلازمة العوز المناعي المكتسب (الايدز AIDS).

المتشابهات النكليوسايدية والعلاج الكيميائي للسرطان

تظهر المتشابهات النكليوسايدية سمية انتقالية ضد الخلايا الخبيثة لكون الاخيرة اكثر حساسية تجاه هذه المركبات من الخلايا السليمة ومنها النكليوسيدات البيرمدينية واليبورينية.

- النكليوسيدات البيرمدينية والبيرمدينات ومنها:

- البريمدينات الفورية ونكليوسيداتها Florinated Pyrimidines وتمثل هذه المجموعة 5 – فلوريوراسل 5-Fura Fluoro Uracil و 5- فلوريوريدين منقوص الاوكسجين -5 Fluoro Deoxyyuridine Faured وقد تم تحضير مجموعة ادوية من fura الفوسفاتية. وهناك 5-ديوكسي-5-فلوريوريدين 5-Deoxy-5-Fluororidine واسترات المحبة للشحوم للـ2-ديوكسي -5-فلورديورين والمركب الحلقي الفوسفاتي الاميديتي -Cyclic Phosphaamidate وكذلك نكليوسيدات الازابيريمدين Azapyrimidine Nucleosides وتتضمن 5-ازاسايتبدين 5-Azacytidine و 5-ازايوريدين 5-Azaurd والنكليوسيدات مزالة الازو Deazanucleosides ومن الامثلة على ذلك 30مزال الازو اليوريديني -3 Deazauridine, 3-Deazauard

النكليوسيدات البيرمديمية محورة السكر

ومن الامثلة على ذلك بيتا (دي-ارابنوفيورانيل سايتوسين) B-D-Arabonfuranosyl Cytosine Arac وقد تم تحضير عدد من الادوية مصدرها الاراك Arac ومنها النكليوتيدات

محددة الطول Oligonucleotides التي تحتوي على الاراك واسترات حامض الكابوكسيل المحب للشحوم.

• المتشابهات النكليوسايدية البيورينية والبيورينات Purines and Purine Nnucleosides ومنها 6-الثايوكولنين (Thio-Gua) 6-Thioguanine و 6- ماركابتوبيورين (6-MP) 6-Merceptopurine وكذلك المتشابهات الادينوسية ومنها فيداربين Vidarbine وفلوداربين Fludarabine.

المتشابهات النكليوسايدية كعلاج كيميائي ضد الفيروسات
Nucleosides Analogues and Antiviral Chemotherapy

ان الهدف الرئيسي للمتشابهات النكليوسايدية الصادرة للفيروس تتمثل بتثبيط الفيروسات وان اغلبية الادوية المضادة للفيروسات التي تم اجازتها عبارة عن متشابهات نكليوسايدية التي يمكن تصنيفها وفق الاتي :

• المتشابهات النكليوسايدية النشطة ضد الفيروسات التي تحتوي على الـ د.ن.أ وتستعمل المتشابهات النكليوسايدية المضادة للفيروسات بصورة رئيسية وتشمل ما يأتي:

– النكليوسايدات البريمدينية والبيورينية ومنها 5-ايودواليوريدين منقوص الاوكسجين 5-idodeoxy uridine (idoxuridine) و 5-تراي فلوروميثيل –2- منقوص الاوكسجين اليوريني 5-trifluoromethyl-2-dideoxyuridine (viroptic) و 3-ازيدو2،6، 3-ثنائية منقوص الاوكسجين الثايمدين (AZT) 3-dideoxythymidine , 3-azido2 والاثيل مزال الاوكسجين اليوريديني Ethyl Deoxyuridine (EDu) و 5-(2-يرمو-ثنائي الايوددفاينيل) مزال اليوريدين (2-bromo and diiodo vinyl deoxyuridine, B4 D4, IVD4) -5 .

• المتشابهات النكليوسايدية النشطة ضد فيروسات الـ د.ن.أ Nucleosides Analogues Active Against Viruses RNA

ان الامراض التنفسية الرئيسة التي تصيب البشر تعود الى الفيروس الذي يحتوي على الـ د.ن.أ ومنها فيروس الانفلونزا اما الفيروسات القهقربة Retroviruse فهو احد الفيروسات ذات

نفس المواصفات والتي تسبب الاسهال. تستعمل الكثير من النكليوسايدات كعلاج ضد الفيروس الذي يحتوي على الـ ر.ن.أ. ومنها 1-بيتا-رييوفيورانسيل-1، 2، 4، ثلاثي الاول-3، كاربوكسايد 1-B – Ribofuranosyl-1, 2,4-triazole-3-Crboxamide اما المتشابهات الادينوسية فتتضمن 3-Deazaadenosine Neplanocin, 3-Deazacarbo Adenosibe وغيرها.

- المتشابهات النكليوسايدية المثبطة للانزيم المستنسخ العكسي reverse transcriptase وتستعمل هذه المتشابهات لعلاج الايدز ومنها:

2,3 dioxy nucleosides (dd Ns)

وتمثل Ziduovudine AzT

Didanosine ddI, (d4T) Stavudine, Lamivudine 3TC, Zalcitabine ddC وهي مجازة لعلاج الايدز.

التطورات الحديثة في التركيب البنائي للبروتينات

- حدثت تطورات كبيرة في التركيب البنائي الثاني للبروتينات ومنها اكتشاف العديد من الحلزونات البنائية المختلفة كالحلزون والشربط وغيرها والتي تختلف والذرات الموجودة في محتوى الاصرة الهيدروجينية عند مقارنتها بالحلزون الفا التقليدي وهذه الاشكال تختلف في نسبها بالبروتينات المختلفة.

- طرق جديدة لاختبار الـ د.ن.أ استحدثت طرق جديدة لاختبار الـ د.ن.أ خلال السنين الماضية ومنها:

- اختبار مجس الـ د.ن.أ.

يوفر استعمال مجسات الاحماض النووية اهمية تفوق الاختبارات المناعية التقليدية، حيث يمكن تحضير كواشف خاصة في ضوءها يحدد تسلسل الجين المراد كشفه وكذلك تشخيص المرض الوراثي. تستعمل مجسات الـ د.ن.أ المتخصصة للكشف عن مختلف الجينات ومقاطع

من الجينات ونواتج الجينات ومنها جينات التوكسينات المتخصصة ودالات المضادات الحياتية المقاومة.

تستعمل مجسات الــ د.ن.أ. بصورة كفوءة بعدد من الطرق التقليدية ومنها:

- التهجين بالترشيح Filter Hybridization
- وصمة سثرن Southern Blot
- وصمة نورثرن Northern Blot
- التهجين بالموضع Insitu Hybridization

- متطلبات وسم الــ د.ن.أ

تتضمن معظم الطرق الحالية المستعملة لوسم وبالتالي الكشف عن مجسات الــ د.ن.أ ، استعمال النظائر المشعة H^3, P^{32}, S^{35} والتي تحسب كميا بواسطة الوميض السائل Liquid Scintillation او بتعريض التقانة الحيوية، الكيمياء الصناعية، والطاقة المتجددة، وادارة النفايات الصلبة ومياه الصرف الصحي، والفيزياء الطبية والاشعاعية، والصحة البيئية والمحميات الطبيعية، والتنوع الاحيائي، والمكافحة الاحيائية لامراض النبات.

يتطلب اعداد رؤية وسياسة وطنية وخطة للسنوات العشر القادمة لنوعية ومستوى التخصصات العلمية التي سوف يكون مطلوبا من قبل الاقتصاد الوطني العراقي، ويفترض ان تشارك في اعداد هذه الرؤية، الجامعات ومراكز البحوث والقطاعات لاستيعاب خريجي هذه التخصصات. واهم هذه القطاعات هي: القطاعات الحكومية-الوزارات والهيئات في مجالات التعليم والصحة والزراعة والصناعة والبيئة والقطاع الخاص ومراكز البحوث والجامعات والمعاهد الفنية التطبيقية، وما الى ذلك. هناك حاجة ملحة لتحديث برامج الجامعة التعليمية من حيث تشجيع ودعم تنمية المعارف والمهارات حدثت تطورات مذهلة في مجال العلوم الكيميائية، وخاصة خلال النصف الثاني من القرن العشرين، بما في ذلك واجهة بديهية وضمنية. يعني التعامل مع التطورات والمحتوى والمفردات والاليات المعروفة في الكيمياء

وتقديم تفسيرات محسنة او جديدة للظواهر والاحداث والتفاعلات الكيميائية تلوث البيئة، والتلوث الضوضائي في معظم انحاء العالم. ويوجد العديد من المنتوجات الاخرى بالاضافة الى كثير من عمليات التقنية التي تؤدي الى تلوث البيئة.

الامن الحيوي وتأثير الاشعة على الانظمة الكيميائية الحياتية

تتميز خلايا الكائن الحي بكونها انظمة كيميائية حياتية بسيطة ومعقدة تتأثر عند تسليط الاشعة المؤينة عليها وبالتالي الى تلف المركبات المساهمة فيها او تغيير نسبها او توليد مركبات اخرى قد لا تستطيع ان تؤدي الوظائف الحيوية التي تقوم بها في جسم الكائن الحي، فقد وجد بالتجربة ان جرعة الاشعة العالية تستطيع ان تستحدث انواعا متعددة من الامراض ومنها الاورام بانواعها من خلال تكوين الجذور الحرة مثلاً ويقوم الشخص الطبيعي باستحداث آلية انزيمية للتخلص منها.

هناك مستويات محددة للامن الحيوي تم وضعها وفق قواعد حددت من قبل مؤسسات (علمية، عالمية) يدخل الاشعاع ضمنها بالتنسيق مع المؤثرات المختلفة الاخرى.

تشكل المواد المشعة المستعملة في عمليات التشخيص والمعالجة والابحاث العلمية بغض النظر عن مصدرها سواءً عبارة عن مستشفى او مؤسسة طبية او بحثية او ناتجة عن الاشعة التشخيصية والعلاجية والطب النووي تهديداً للصحة العامة والبيئة وهي نفس الوقت يد الطبيب اليمنى في معركته ضد المرض.

تستخدم النظائر المشعة مثلا في ميدان التشخيص الشعاعي كمشخص للدواء كما وتجري عملية الكشف والتشخيص بالنظائر المشعة في خطوات محددة ومراحل متتابعة واصبح بالامكان دراسة جميع الاعضاء في الجسم بالنظائر المشعة كالدماغ والغدة الدرقية والقلب والرئتين والكبد والطحال والكليتين وغيرها.

ومع ان كمية النظائر المشعة المرتبطة بالعدد التشخيصية التي تستعمل في الاختبار المناعي الشعاعي صغيرة، لكنها قد تسبب مضاعفات بالنظير المشع I^{131}, I^{121} مثلا في حالة تناولها

يتكسر المركب النشط اشعاعيا عندئذ في الجسم ويمتص انثيود المشع ويركز في الغدة الدرقية، لذا فالغدة الدرقية تستلم جرع اكبر من الاشعاع المسموح به نتيجة التوزيع العشوائي في الجسم، فعليه وضع ضوابط للامن الحيوي عند استعمال هذه النظائر المشعة بالرغم من صغر الكميات المشعة المستعملة تشكل الكيمياويات عموما والنظائر المشعة وغيرها مستويات عدة من الامن الحيوي يمكن الاشارة اليها خلال البحث.

بعض الاتجاهات الحديثة في الرياضيات

ذهبت بعض الجامعات الى استحداث مشروع Project وعيادات رياضية Mathematical Clinics لتأكيد التطبيقات، فضلاً عن ذلك فقد استحدثت دراسات بينية مثل رياضيات واحصاء، رياضيات وفلسفة، رياضيات واقتصاد، رياضيات وفيزياء، رياضيات وهندسة، رياضيات وحاسبات، وغيرها. وادخلت الحاسبات وسيلة لتدريس بعض مواضيع الرياضيات واستنباط طرق تدريس جديدة بديلا عن المحاضرة السائدة حاليا والاهتمام بنشر الثقافة الرياضياتية والوعي الرياضياتي. تم التأكيد على وحدة الرياضيات كفلسفة، باتجاهات متعددة (محاور) مثل رياضيات تطبيقية (هندسية)، علوم حياة رياضية، اقتصاد رياضي، رياضيات مالية، احتمالية، احصاء، بحوث عمليات، الرياضيات البحثية وغيرها. كما تم تأكيد تطبيقات الرياضيات خلال البرامج العامة، وبصورة خاصة اعطاء اهمية خاصة للمنذجة الرياضية، وقد ذهبت بعض الجامعات الى استحداث مشروع Project وعيادات رياضية Mathematical Clinics لتأكيد التطبيقات. فضلاً عن ذلك فقد استحدثت دراسات بينية مثل رياضيات واحصاء، رياضيات وفلسفة، رياضيات واقتصاد، رياضيات وفيزياء، رياضيات وهندسة، رياضيات وحاسبات، وغيرها. وادخلت الحاسبات وسيلة لتدريس بعض مواضيع الرياضيات واستنباط طرق تدريس جديدة بديلا عن المحاضرة السائدة حالياً والاهتمام بنشر الثقافة الرياضياتية والوعي الرياضياتي: للخربجين.

تم التأكيد على وحدة الرياضيات كفلسفة وتسمية الاقسام المعنية بالرياضيات باسم قسم العلوم الرياضياتية ويضم هذا القسم اتجاهات متعددة (محاور) مثل رياضيات تطبيقية (هندسية)، علوم حياة رياضية، اقتصاد رياضي، رياضيات مالية، احتمالية واحصاء، بحوث عمليات، الرياضيات البحثية وغيرها. كما تم التأكيد على تطبيقات الرياضيات خلال البرامج العامة، وبصورة خاصة اعطاء اهمية خاصة للمنذجة الرياضياتية، وقد ذهبت بعض الجامعات الى استحداث مشروع (Project) وعيادات رياضياتية (Mathematical Clinics) للتاكيد على التطبيقات فضلاً عن ذلك فقد استحدثت دراسات بينية مثل رياضيات واحصاء، رياضيات وفلسفة، رياضيات واقتصاد، رياضيات وفيزياء، رياضيات وهندسة، رياضيات وحاسبات، وغيرها. وادخلت الحاسبات كوسيلة لتدريس بعض مواضيع الرياضيات واستنباط طرق تدريس جديدة بديلا عن المحاضرة السائدة حالياً والاهتمام بنشر الثقافة الرياضياتية والوعي الرياضياتي. واخيراً تطمح اقسام الرياضيات اليوم لتزويد الطلبة بمؤهلات يمكن التعبير عنها بما يأتي :-

- ثقافة كافية تؤهلهم لفهم الصعوبات.

- ثقة كافية بالنفس تشجعهم على معالجة قضايا فنية صعبة.

- مقدار من الشك يجعلهم يسألون الاسئلة الصحيحة.

- مقدار من المثابرة يجعلهم يواصلون التفتيش عن اجوبة مناسبة.

-

- حسن التقدير لاختيار ما هو صحيح.

الاختصاصات المستقبلية في العلوم الصرفة

يعرف العلم Science بـ (مجموعة المعارف والحقائق والخبرات الانسانية) يشمل العلوم التطبيقية وهي -بدقة-ترجمة للعبارة الانكليزية (National Scince) او (Science of

419

Nature) ويعني بها المعلومة التي تختص بدراسة الطبيعة التي تحيط بالانسان من احياء وجمادات وكل ما يتعلق بالارض والجو والاجرام السماوية. ويمكن ان تنقسم الى قسمين: علوم اساسية وعلوم تطبيقية، فالعلوم الاساسية عند البعض جميع العلوم الطبيعية ما عدا الهندسة والعلوم التطبيقية هي الهندسة بتطبيقاتها المختلفة.

في حين يعتقد البعض الاخر ان العلوم الاساسية المسماة بالصرفة تشمل الرياضيات والفيزياء والكيمياء وعلوم الحياة وعلوم الارض والفلك والارصاد الجوية وهو ما نقصده في هذه الدراسة.

يتميز العصر الذي نعيش فيه بتفجر العلوم الصرفة بسرعة مذهلة، وما ان تبتدع عملية جديدة او تعرض فكرة جديدة، حتى يهرع التكنولوجيون الى استغلالها بابتكار جديد. فالحاسوب مثلا كان قبل سنوات موضوعا يتكلم عنه الجامعيون صار اليوم الشغل الشاغل في الدوائر والمؤسسات والمصانع. ان التكيف للتفجر العلمي الصرف يفترض تغييراً جذرياً في مفاهيم التعلم والتعليم والمناهج التي تقتضيها الانتقال الى المرحلة الحضارية المتقدمة التي ننشدها، لا يمكن ان تنهض بها الجامعة فقط بل المؤسسات ذات العلاقة.

ان تفجر العلوم الصرفة يفرض على الجامعات ان تقوم بالقسط الاوفى بحيث تلائم الحاجات المحلية فعلى الجامعات بخاصة ان تضع برامج تعليمية ومناهج تفيد من البرامج المستوردة، لاننا في عالم سريع التطور يكاد كل يوم يشمل تطورا علميا وتقنيا جديدا، ينبغي مثلا ان يتوافر في الجامعات والمؤسسات وسائل فعالة للحصول على المعلومات ونشرها وان تحديات العصر تفرض على هذه الجامعات ان يكون لها ملاكات خاصة.

تصورات مستقبلية

• يتصف علم الانسان بمحدودية ونسبية وعدم مطلقية معارفه بحكم ارتباطها بزمن ومكان محدودين تقتضي ان يكون غير شامل وغير ثابت وغير ملم بالمستقبل وانه قابل

للتغير والتطور والتعديل، وان ما يتوصل الانسان من معارف وقوانين علمية في زمن معين قد لا يصبح صحيحاً ومقبولاً في زمن لاحق. وان علم الانسان وان ألم ببعض ما وجد بالفعل فانه عاجز عن الالمام بما يحدث في المستقبل وان العلم البشري يتقدم مع الزمن، وان ما لم يكتشفه العقل البشري اليوم قد يكتشفه غدا، وان العقل البشري لا يكتشف ولا يدرك حقائق الكون دفعة واحدة بل في مراحل مختلفة من الحياة وان السعي في سبيل المعرفة عملية مستمرة. ولكن رغم محدودية عقل الانسان وعجزه في كثير من الاحيان عن فهم العديد من القضايا العلمية، وتفسيرها فهو دائم السعي لفهم الغازها ولم تمنعه من البحث عن المستقبل بعد ما اصبحت الادوات اللازمة بدراسة المستقبل والتنبؤ به متوفرة. واصبحنا نرى التقدم العلمي في الكيمياء وعلوم الحياة والرياضيات وعلوم الحاسبات والعلوم الاخرى يتزايد سنوياً واصبحت اساساً لمعظم العلوم الاخرى.

- من الممكن ان نقسم الدراسات المستقبلية في العلوم الى ثلاثة انواع:

- الدراسات التي تعتمد على التنبؤ.

- الدراسات المستقبلية التي تعتمد على الحدس.

- الدراسات المستقبلية التي تعتمد على معلومات احصائية مفصلة ضمن برامج رياضية او نماذج حاسوبية.

وفي ضوء ذلك هناك اربعة تصورات يمكن الاشارة اليها للمستقبل وهي:

- تصورات شديدة التشاؤم.

- تصورات متشائمة حذرة.

- تصورات متفائلة حذرة.

- تصورات متحمسة للنمو والتكنولوجيا.

وان تحتاج هذه الدراسات الى نماذج مستقبلية تتطلب العديد من الامور منها:

- قاعدة معلومات تشمل احصاءات شاملة.

421

- مجموعة من الدراسات الاولية التي تعتمد على الاسقاطات سواء بالنسبة للسكان ام الموارد.

- اعتماد اهداف محددة.

- حاسبات ذات قدرة كبيرة على الخزن والتعامل مع المعلومات.

علوم الحياة

تشير التوجهات الحديثة في علوم الحياة الى التطورات التي حصلت في تخصصات علوم الحياة وهي من اكثر جوانب المعرفة نمواً واتساعاً في عصرنا هذا. ان هذا التوسع الهائل في جوانب المعرفة الذي تحقق خلال النصف الثاني من القرن العشرين يمكن وصفه بانه انفجاري، اذ يصعب على المرء مواكبة ما يصدر ويوثق عن تلك الاكتشافات التي تحول العلوم من طبيعتها الوصفية للاحياء الى علوم قادرة على ان تحدث تغييراً في الاحياء وفي اشكال الحياة وانماطها وبيئتها مما احدث ازدياداً في الاهتمام العالمي بها وبتأثيراتها على الحضارة البشرية عموماً. ومن المتوقع ان يشهد القرن الحادي والعشرون مزيداً من هذه التطورات حتى ليعتقد البعض ان التطورات في علوم الحياة ستكون سمة القرن الحادي والعشرين واحدى دعامات العلم والثقافة التي ستركز على حضارته، كما ان المخاطر التي ستنتجم عن تطبيقاته ستكون اهم التحديات التي سيواجهها الانسان في المستقبل القرب.

علوم الحياة

فيما يأتي نورد بعضاً من اهم التخصصات المستقبلية في علوم الحياة، ومن الجدير بالذكر ان لا نتناول التخصصات الموجودة في القطر حالياً والتي تم انشاؤها اثناء انشاء الهياكل الاكاديمية والبحثية التابعة لها.

-علوم البيئة.

-علم حياة الخلية Cell Biology

علم المناعة Immunology

-علم السموم Toxicology

-علم بيولوجية الاعصاب Neurobiology

-العلوم الاحيائية الطبية Biomedical Science

-الوراثة البشرية Humman Genetics

-علم الحياة الجزيئي Molecular Biology

-الكيمياء الحيوية Biochemistry

الاخلاقيات الاحيائية Bioemics

وكذلك بعض الاقسام المقترحة في ضوء التطورات العالمية:-

– قسم العلوم الوراثية.

– قسم الكيمياء الحيوية.

– قسم المناعة.

– قسم علم حياة الخلية.

المسارات والتخصات لكل قسم

ان لكل قسم مسارات رئيسة يمكن ان تستنبط منها المساقات الدراسية للدراسات الاولية والعيا وحسب المرحلة التي يمر بها القسم المعني، كما تحدد بضوئها الخطوط البحثية والتخصصات داخل ذلك القسم ونورد فيما يأني بعضاً منها:

قسم البيئة

ويشمل المسارات التالية:

-تلوث البيئة الزراعية.

-تلوث البراري والصحارى.

-تلوث المدن.

-تلوث المياه.

-آثار المشاريع التنموية على البيئة.

-التنوع الاحيائي Biological Diversity

-التصحر.

-المكافحة الاحيائية Biological Control

-بدائل مضبات الاوزون.

-بدائل النظائر المشعة الاحيائية.

-علم الحياة المناخي Enviromental Biology

دور الملاكات العلمية:

-مما لاشك فيه ان دور الملاكات العلمية المتخصصة العاملة في الجامعات هو دور قيادي وفعال ولتمكين هذه الكوادر اداء هذا يتعين تحديد ما هو مطلوب منها وهو :- التدريس.

-القيام بالدراسات النظرية الاساسية الميدانية لوضع الخطط البحثية المستقبلية.

-اجراء البحوث، و توجيهها.

-القيام بالاستشارات العلمية والفنية من خلال المكاتب الاستشارية او بطرق اخرى.

ولتمكين اعضاء الهيئة التدريسية من القيام بهذه المهام يجب اعادة النظر بهيكل عمل التدريسي.

فلقد تعرض الدور الاساسي الذي يفترض ان يقوم به الاستاذ الجامعي والمتمثل في الموازنة.

المراكز البحثية الحديثة في ضوء البحوث المستقبلية للعلوم الحياتية المقترح تأسيسها:

424

مراكز البحوث

-اعادة تأسيس بحوث النخيل.

-انشاء (بنك) لحفظ وحماية الاصول الوراثية والبذور العراقية.

-انشاء مركز لاستخدامات الحاسوب في علوم الحياة.

-انشاء المحميات الطبيعية الجديدة.

-تأسيس مركز بحوث متخصص لحصر الاثار الصحية والبيئية التي سببها الحصار الظالم للتمكن من تحديدها والقضاء على آثارها.

-انشاء مركز وطني لبحوث البيئة والتنوع الاحيائي وحماية الاصول الوراثية الوطنية، خاصة تلك المتعلقة بالمحاصيل الستراتيجية.

استثمار التوجهات الحديثة في علوم الحياة

– التركيز على استكمال الدراسات الهادفة الى تعرف المناطق الجغرافية الاحتياطية في العراق والبيئات المختلفة والتنوع الاحيائي في كل منها لغرض اعداد مسح كامل للثروة الحياتية في العراق.

– التركيو على الثروات الحياتية التي يتفرد بها العراق كالنخيل والجمال والجاموس والماعز والقصب والكمأة.

– دراسات البيئة المائية والثروة السمكية.

– دراسة بيئة الاهوار.

– دراسات الحد من التصحر.

– دراسة المشاكل الاحيائية في الصناعة والتصنيع.

– بحوث ودراسات لتنظيف البيئة من التلوث.

– دراسات وبحوث عن الامراض المتوارثة في العراق، خاصة الاسر المعتمدة على الزواج من الاقارب.

علوم الفيزياء

ان التوجهات الحالية في اقسام الفيزياء في العديد من الجامعات الاجنبية الرائدة تتضمن التعددية في مفهوم شهادة البكالوريوس فيزياء حيث يتم منح تلك الشهادة بمثابة موضوع موحد يضم مفردات الفيزياء الاساسية كالنووية والصلبة والذرية زيادة على دروس في الرياضيات. او يتم منح شهادة البكالوريوس في محور معين ضمن التخصص مثل فيزياء الفلك او فيزياء البيئة وهذا يضم بعض مفردات الفيزياء الاساسية مزيداً عليه مواضيع المحور الشخصي المحدد، واخيراً يتم منح شهادة البكالوريوس في دراسة بينية (او مشتركة) تضم بعض مفردات الفيزياء الاساسية مزيدا عليها مفردات الموضوع الثاني والامثلة على ذلك تشمل الفيزياء الرياضية او الفيزياء الكيميائية او الفيزياء والهندسة الالكترونية او الفيزياء وعلم الحاسبات... الخ. ان حصيلة هذه التعددية يمكن ان ترفد المجتمع بتنوع غني ومتكامل وان يلبي الحاجة الانية في الصناعة او غيرها كما يمكن ان يبني الملاك الجامعي الذي يقود العملية البحثية.

تحديد الاختصاصات المستقبلية في الفيزياء

– بالنسبة للدراسة التي تعتمد محاور معينة في ضمن التخصص الواحد فانه يمكن توسيع قاعدتها باضافة مواضيع اختيارية او ربما استحداث محاور معينة مثل فيزياء التقانات الحديثة وفيزياء البيئة واخرى اشير اليها.

– بالنسبة للدراسات البينية التي يجري العمل فيها بشكل محدود (الجامعة التكنلوجية) فانها وفي مرحلة الدراسة الجامعية استحداث خطة لتحديد الاهداف حتى يتسنى وضع مناهج متكاملة، فدراسة مشتركة بين الفيزياء وعلوم الحياة او الطب يمكن ان تقدم الكثير لصناعة الاطراف الناعية مثلا وتقوم بعض اقسام الهندسة بدور الفيزياء في هذا المجال فيما يعرف بـ Biomedical Engineering وقد حل الفيزياء في ميدان الالعاب الرياضية بشكل اساسي كما تشترك الفيزياء والهندسة في مجال الفيزياء الالكترونية او الهندسة الالكترونية.

426

- تبقى المسألة الاساس في كل العملية التعليمية وهي مسألة الكادر التدريسي ووجوب بقائه على اتصال مع آخر التطورات فعلم الفيزياء يتطور بسرعة مذهلة وتتعدد مجالات تطبيقاته كما ان محددات دراسة الدكتوراه لتوفير النقص في الملاك يمكن معالجتها وهي محددات مادية في الغالب تشمل تحديث اجهزة المختبرات وطرائق الاتصال بالعالم، هذه الامور ترتقي بالنوع وهو الامر المطلوب دائما.

علوم الارض (جيولوجيا)

اعتمد مبدأ التخصص في فروع الارض المختلفة مع المحافظة على المناهج الاساسية لعلوم الصرفة المشتركة. التجارب العالمية في هذا المجال متنوعة (على مستوى البكالوريوس)، تتراوح من التخصص العام (جيوبوجي عام) الى التخصص الدقيق جداً الموجود في بعض الجامعات ذات الطبيعة الخاصة (مثلا جيولوجيا التعدين في بعض الجامعات مثل كاردف، الامبريال، كولورادو)، غير ان التطور الحديث هو ان تشمل الاقسام واحداً او بعضاً من اربعة تخصصات رئيسة في ضمن علم الارض هي:

- الجيولوجيا والصخور والمعادن.

- الجيوفيزياء.

- المتحجرات.

- المياه الجوفية والجيولوجيا البيئية.

تحديد الاختصاصات المستقبلية:

- المواضيع الدراسية.

- الفروع والاقسام العلمية.

- الدراسات العليا.

- فتح المراكز البحثية.

427

▪ دور الملاكات.

اهمية الاختصاصات المزمع استحداثها.

المواضيع الدراسية في المراحل الاولية:

- تحديث مفردات المناهج بشكل عام للمواضيع الاساسية لعلوم الارض واعتماد الكتب الحديثة في ذلك.

- تدريس المواضيع الحديثة في ضمن دروس الاختصاص مثل : جيولوجيا البيئة، جيوفيزياء هندسية، برامجيات النمذجة الرياضية، جيولوجيا منجمية، الطباقية الزلزالية، طباقية التتابع، تحليل الشحنات الرسوبية، جيولوجيا البحار، جيولوجيا الاثار والعصر الرباعي، علم التربة، جيوكيمياء عضوية، التطبيقات الحديثة في التحسس النائي، استكشاف معدني، جس بئري، فيزياء بنيوية، هندسة المعادن.

- الاقسام العلمية المتخصصة المقترحة/ جيولوجيا النفط، جيولوجيا الموارد المعدنية والصخور الصناعية، جيولوجيا الموارد المائية، الجيولوجيا البيئية والهندسية، جيولوجيا البحار.

- الدراسات العليا: تعزيز المناهج التطبيقية في الدراسات العليا وتأكيد المواضيع ذات العلاقة بالموارد المعدنية والنفطية والمائية اضافة الى التخصصات العليا الصرفة.

- فتح مراكز بحثية حديثة : مركز بحوث المياه، مركز بحوث النفط، مركز بحوث التصحر، مركز بحوث الجيولوجيا الهندسية، مركز بحوث العصر الرباعي وجيولوجيا الاثار.

▪

الكيمياء

عند مقارنة مناهج اقسام الكيمياء في الجامعات العراقية بما هو جار عالمياً يتضح لنا ما يأتي:-

- ظلت مناهج الكيمياء في جامعاتنا محافظة على اطرها العامة التي رسمت لها منذ بداية السبعينيات.

- تتصف مناهج اقسام الكيمياء العراقية بمحدودية للموضوعات المقررة وبمحدودية المفردات والمضامين.

- خلو المناهج الدراسية لاقسام الكيمياء وجامعاتنا من الموضوعات التخصصية الحديثة (الضمنية والبينية على السواء) ومن التقنيات الكثيرة التي استجدت.

- التقنيات والطرائق الطيفية والمجهرية الحديثة واستخدامها في الكيمياء لم تدخل في الموضوعات والمناهج الدراسية.

- الموضوعات المساعدة الحديثة في الكيمياء تكاد تكون غائبة برغم اهمية تلك الموضوعات في مناهج الكيمياء الحديثة.

- ان موضوعات الحاسوب التخصصية غائبة كليا عن مناهجنا المقررة.

- لم تدخل الرموز والمختصرات والتسميات الكيميائية الدولية المعمول بها في ميدان الكيمياء منذ اوائل السبعينيات.

- التفاعلات الكيميائية النووية بشكل خاص والاشعاعية بشكل عام لا يتم في موضوعات الكيمياء المختلفة الموجودة في ضمن مناهج اقسام الكيمياء في جامعاتنا.

ظلت هياكل كليات العلوم العراقية بعيدة عن التغيرات الجوهرية على الرغم من استحداث بعض الاقسام العلمية الجديدة في البعض منها، وتكاد تكون كليات العلوم نسخاً متشابهة بعضها لبعض من حيث محتواها من الوحدات العلمية والمناهج الدراسية والدراسات العليا. ولم تجر محاولات جادة لحد الان لاستحداث كليات علوم تطبيقية او انشاء اقسام تطبيقية بجانب الصرفة منها، او العمل على شمول مناهج الاقسام العلمية في كليات العلوم العراقية بالجنبين الصرف والتطبيقي معاً.

ان العلوم الضمنية والبيئية الكثيرة التي أظهرت في كل اختصاص رئيسي في العقود الاخيرة من هذا القرن اصبحت تستدعي انشاء فروع اقسام او كليات جديدة واستحداث قنوات جديدة للدراسات العليا والبحث العلمي.

وبات ضرورياً منح شهادات جامعية عليا واولية في الكثير من تلك الاختصاصات. ووفق ذلك نشير الى بعض الاختصاصات التي يمكن ان تؤلّف اقساماً ضمن كليات العلوم الحالية في الجامعات او ادخالها في ضمن كليات العلوم التطبيقية التي سيتم انشاؤها مستقبلاً ومنها:-

- الكيمياء الحياتية.

- الكيمياء التطبيقية.

- كيمياء النفط والكيميائيات النفطية.

- علوم التآكل.

- علوم السطح.

- العمل المساعد والعوامل المساعدة.

- علم الحرائق.

- علوم الجزيئات الكبيرة ومتعددات الجزيئات (البوليمرات).

- علم المواد.

- الفيزياء الكيميائية.

الرياضيات

وفق التطورات في منهجية الرياضيات:-

- يمكن زيادة بعض المواضيع الاختيارية، واستحداث محاور معينة في بعض اقسام الرياضيات التي تسمح امكاناتها بذلك مثل محور احصاء رياضي، رياضيات هندسية، رياضيات حسابية وغيرها.

- استحداث دراسات بينية مع اقسام اخرى مثل رياضيات وحاسبات، وفيزياء، رياضيات وعلوم حياة، رياضيات وفلسفة وغيرها. وحسب ما تسمح به ظروف القسم وامكاناته. ان التداخل بين الرياضيات والعلوم الاخرى هو السائد اليوم و نعتقد بانه سيستمر في المستقبل.

- ان الاختصاصات التي يحتاجها العراق –على مستوى الدراسات العليا– تشمل الاختصاصات المعروفة عالميا للرياضيات كافة.

الحاسوب

يمكن ان نستنتج ان اهم التخصصات على المدى المستقبلي القريب تتمثل في المواضيع الاتية:–

بالنسبة للمكونات المادية للحاسوب سيكون هناك مواضيع مثل تقنيات الحاسوب، الروبوتات، اما اما على مستوى البرامجيات ستكون المواضيع هي نظم المعلومات وهندسة النظم وهندسة البرامجيات وتقنيات المعلومات والمعلوماتية والانظمة الخبيرة والذكاء الاصطناعي والشبكات العصبية ونظم دعم القرار والترجمة الالية والالسينات الحاسوبية والواقع الافتراضي والقراءة الضوئية وشبكات المعلوماتية وامن المعلومات والحواسيب والتشفير بالاضافة الى علم الحاسوب وعلم معالجة البيانات. وعلوم النظم المعرفية. ونظم التعليم والتعلم بواسطة الحاسوب. ونظم التصميم والانتاج بواسطة الحاسوب والشبكات العصبية. والنظم المعرفية.

وان مثل هذه التخصصات تصلح ان تكون كليات مستقلة ككلية علوم النظم المعرفية او اقسام مثل هندسة النظم وهندسة البرامجيات ونظم دعم القرار او تكنولوجيا الحاسوب. اما المواضيع التي يمكن ان تكون مجال او عنوان لدراسات عليا كالذكاء الاصطناعي، الروبوتات، الانظمة الخبيرة، شبكة المعلومات، وامن المعلومات والحواسيب، والترجمة الالية، والقراءة الضوئية واللسانيات الحاسوبية وتعليم الحاسوب والواقع الافتراضي. اما المواضيع التي تتطلب

431

مراكز بحثية متخصصة هي مركز المعلومات ومراكز بحثية في مجال القراءة الضوئية وتحليل الصوت وتركيبه والروبوتات ومركز CAM/CAD ومراكز الصناعة البرمجية وغيرها.

وفي ضوء ذلك يمكن القيام بما يأتي:-

– فتح اقسام علمية تخصصية في مجال الحاسوب مثل اقسام تقنيات الحاسوب والروبوتات، هندسة النظم، الذكاء الاصطناعي والمعلوماتية وادارة وامن الحواسيب، ونظم المعلومات.

– فتح دراسات عليا متخصصة في المجالات الاتية:

الالسينات الحاسوبية، والواقع الافتراضي، ونظم دعم القرار، والترجمة الالية والشبكات العصبية والتفجير، وشبكات المعلومات وحمايتها.

– فتح كلية عملية للحاسوب تحتوي على اقسام حاسوبية مختلفة تشمل النظم المعرفية والمعلوماتية ونظم المعلومات وامن الحاسبات والمعلومات.

– اعطاء اهمية خاصة لفتح كليات متخصصة لاستخدام الحاسوب في التعليم والتعلم بدلا من اقسام الحاسوب في كليات التربية.

– تأسيس مراكز بحثية ومراكز متخصصة اخرى وكما يأتي :

أ-مركز متخصص للاهتمام باجراء بحوث اللغة العربية والحاسوب ومن جوانب الكلام والكتابة والترجمة كافة او لغات البحث او الاسترجاع او برامجيات تشغيلية عربية.

ب-مراكز متخصصة في مجال دعم الصناعة البرمجية.

ج-مركز متخصص في استخدامات الحاسوب في التصميم والانتاج CAM/CAD.

د-مركز وطني للمعلومات.

– التركيز في الدراسات العليا لاعداد ملاكات علمية تدريسية في التخصصات المستقبلية المقترحة وذلك بهدف اعداد التدريسيين اللازمين لانشاء اقسام علمية جديدة.

مقترحات وتوصيات عامة

-استحداث الدراسات البينية مع توافر امكاناتها في مرحلة الدراسات الاولية تمتلك اختصاصا رئيسا وفرعيا ويرسل الى الخارج مستقبلا للحصول على شهادة عليا في الاختصاص المعني.

-انشاء مركز متخصص بالرياضيات يعمل فيه اساتذة اجانب والخبرة الاجنبية او ارسال بعثات خارج العراق تعتمد على الاستعانة بالموارد التي يدفعها الطلبة العرب.

-توفر مستوى معاشي لائق للتدريس يجعل منه عنصراً علمياً متفرغا يايح له الوقت الكافي للتأمل والاطلاع والتعويض عن الاساتذة.

-استحداث هيئة قطاعية لعلم الحاسوب في وزارة التعليم العالي لوضع المناهج.

-استحداث دراسات عليا في العوامل الذكية وفتح مراكز بحثية لصناعة البرامجيات.

-ارسال بعثات بتخصصات جديدة ومستحدثة.

-قيام لجان خارجية من الدول المتقدمة لتجديد المناهج.

-تطوير الدراسات العليا في داخل القطر للتعويض عن الصعوبات التي يواجهها طلبتنا في الحصول على قبول خارج القطر.

-استخدام خبراء من الدول المتقدمة لتدريب العاملين في القطر.

-دراسة شاملة للتعليم (الابتدائي، الثانوي، الجامعي) لخلق الانسان المبدع والمفكر، الاساسيات الاولية والدقيقة للدراسات العليا.

-نغيير النظام الجامعي واعادة النظر بشكله جذرياً والاعتماد على تجربة 1992 (نظام الوحدات والمقررات).

-القيام بعقد مع الجامعات المتقدمة لتحديث التعليم الجامعي في القطر.

-القيام بدراسات جديدة تحمل ارقاما محددة.

-تقويم جديد للمناهج الدراسية المعمول بها.

-تقويم لسياسة القبول في الدراسات العليا.

-تقويم للتطورات العلمية العالمية وموقفنا منها.

-تقويم المستوى لخريجي الدراسات العليا (الماجستير والدكتوراه) ومدى اهليتهم للدراسات العليا.

-استحداث برنامج وطني لتطوير خريجي الدراسات العليا في القطر (الماجستير والدكتوراه).

-وضع موضوع الحاسوب كموضوع اساسي في الدراسات العليا.

-تقويم لحالة عزوف الطلبة من الدخول في بعض التخصصات (علم الارض مثلا).

تأكيد التخصص العام في الدراسات الاولية والدقيقة في الدراسات العليا

العلوم الانسانية

تنقسم العلوم الى قسمين بصورة عامة من العلوم الانسانية والصرفة، فالانسانية والاجتماعية تخص لجميع ما يتعلق بالانسان وما ينتج من سلوكيات ومواقف وبالتالي فان العلوم الانسانية تتناول الانسان والمجتمع والانسان والطبيعة والانسان.

اما المدارس التي تناولت فهم الانسان فعديدة منها المدرسة اليونانية فتناولت اهتمام اليونانيين من امثال سقراط وارسطو وافلاطون بالانسان وكانت اجتهادات هؤلاء الفلاسفة تشير الى طريقة تحليله فكانت ترتكز على الراي العقلي اكبر من الاي التطبيقي والعلمي ، اما موقع الانسان في الحضارة الرومانية فتعتبر حالة استمرارية لعلوم اليونان في الادب والفكر.

اما النظريات اللاحقة فكثيرة فمنها الماركسية والمدارس الاخرى المعروفة في العلوم الاجتماعية فالباحث لا يجب ان يفكر بالمفاهيم والمفردات واساليب التحليل لما يدرس ويحلل، لهذا كان منطقياً ان يفكر من داخل تصوراته العلمية الذي يعطى للظاهرة العلمية معانيها وللظواهر العلمية .

434

ووفق ما ذكرناه فان الانسان وفق كافة الاجتهادات سلبا وايجابا هو الانسان في تفاعله مع الاخر يجب ان ينظر الى هذه العلاقة نظرة منهجية شاملة متكاملة في علاقاتها ويتطلب فكر منطلق في جوهر الانسان وعقله وفكره وفق التوازن والايجابية فالثنائية المزدوجة تبحث الانسان في تناسقه وتنوعه.

ساهمت المنجزات العربية الاسلامية والنتاج الصيني والحساب الهندي والخطاب الروماني والحوارات اليونانية وكذلك القصص الفارسية في نهضة العلوم فضلاً عن النظام التعليمي السومري قبل عام (3500-ق.م) واكتشافهم نظام الكتابة.

لعبت الحضارة المصرية ادواراً مهمة في التقدم الذي حصل بدءً من حوالي (300 سنة ق.م) ويعتقد البعض ان المصريين الاوائل هم مصدر الاشعاع التربوي وان الحضارة المصرية هي التي صنعت الحضارة الاغريقية وان التعليم كان مقره غالبا المعابد (عين شمس ومعبد الكرنك) وكان التدريس يتم على خشب الابنوس والعاج وصفحات ورق البردي. ويعتبر نظاما للكتاب بنجاح الخط الهيروغليفي كما يعتقد الاخرون ان مفهوم التربية في اطاره العلمي المعاصر قد تم من قبل المصريين القدماء وان الوسائل التعليمية كانت قد تمت من قبلهم بالرسومات المصرية القديمة المنقوشة على واجهة الجدران بالمعابد كوسيلة من الوسائل الحسية لتوصيل المعلومات.

اما التاريخ اليوناني فيتمثل بمدينتين هامتين (اسبرطة واثينا) اللتين اتفقتا على جعل التربية وسيلة لاعداد المواطن الذي يخدم بلده، ففي اسبرطة كانت الاستراتيجية التعليمية تركز على التدريبات العسكرية المستمرة وبعلماء متميزين اثروا التاريخ بمنجزاتهم من ابو قراط ابو الطب الذي نهض به وناقش المسائل الطبية نظريا وعمليا ولازال قسم ابو قراط Hippocratic Oath من اقدم الوثائق التاريخية النادرة التي تبين اخلاقيات ممارسة مهنة الطب. فاليونانيون من اوائل الناس الذين شرعوا في دراسة الذاكرة ووظائف كل من الدماغ والقلب.

اما الحضارة الرومانية فقد لعبت دورا في تاريخ الفكر البشري، حيث كان التعليم يركز على مهارة القراءة والكتابة وتعلم الصغار في المدارس، وكانت المدارس الابتدائية معروفة باسم اللودوس ويسمى المعلم بالمؤدب ولازالت الافكار التربوية بارزة كونها توجيهات تربوية.

اما الصين وحضارتها فكانت ابعد الحضارات عهدا بالتربية واشهرها ذكرا في التاريخ وكانت المذاهب الفلسفية الصينية تتمثل بالكونفوشيوسية منذ القرن السادس قبل الميلاد. فالحضارة الصينية متمثلة في علوم كثيرة منها الطب والكيمياء ولازالت الابر الصينية وسيلة من وسائل العلاج.

للصين تربية شرقية تتسم بالصرامة والمتمثل بالتقليد وتقدير الماضي والالتزام بسلسلة من التعاليم والحكم الانسانية بالتقليد وتقدير الماضي وتركزت على:

- عقد الاختبارات للطلاب.

- حفظ المعلومات.

- الروحانيات.

- الالتزام بالقوانين والتشريعات.

اما كونفوشيوس (479 ق.م-55 ق.م) وهو من اسس الحركة الكونفوشيوسية، حيث استخدم عدة وسائل للتنشئة الاجتماعية ومنها:

- الاستفادة من المناصب السياسية.

- الاعتناء بمهارة مطالعة الكتب.

- دراسة اللغة والادب.

- طاعة الوالد والخضوع له.

- طاعة الحاكم.

وعليه فان التربية الصينية تتميز بـ:

- التربية الاخلاقية.

- وضوح الرؤيا.

- الايمان بالتربية المستمرة.

- التركيز على تراث السابقين.

اما الحضارة الهندية فمتنوعة، فمنها البوذية التي تتميز بالتفرغ للعبادة وتصفية النفس وان كل الاشياء تحوطها الصعاب والشهوات من اسباب المعاناة. وقد احرزت تعاليم بوذا شهرة واسعة على مستوى العالم، اما الملامح التعليمية عند بوذا فتمثلت بوسائله في نشر التعليم وتعاليمه عن طريق المحاورة والمحاضرة، اما وصاياه فتتركز على عدم القتل والزنا والسرقة والمسكرات. وتناولت الحضارة الهندية الفلسفة والطب ودراسة النجوم والعلوم والرياضيات. ومن النوادر الثقافية كتاب (كليلة ودمنة) وهو يهتم باصلاح النفوس وضعه الفيلسوف الهندي البيدبا. اما الملامح التربوية الهندية فتتمثل بالتربية الروحية والاخلاقية والاجتماعية.

تصورات مستقبلية

يتصف علم الانسان بمحدودية ونسبية وعدم مطلقية معارفه تقتضي ان يكون غير شامل وغير ثابت وانه قابل للتغير والتطور والتعديل، وان ما يتوصل الانسان من معارف وقوانين علمية في زمن معين قد لا يصبح صحيحاً ومقبولاً في زمن لاحق. وان العلم البشري يتقدم مع الزمن، وان العقل البشري لا يكتشف ولا يدرك حقائق الكون دفعة واحدة بل في مراحل مختلفة من الحياة وان السعي في سبيل المعرفة عملية مستمرة. ولكن رغم محدودية عقل الانسان وعجزه في كثير من الاحيان عن فهم وتفسير العديد من القضايا العلمية، فهو دائم السعي لفهم الغازها ولم تمنعه من البحث عن المستقبل بعدما اصبحت الادوات اللازمة بدراسة واستشراق المستقبل والتنبؤية متوفرة.

لعل ما يميز الانسان عن بقية المخلوقات كونه مخلوقا مستقبليا، فمنذ العصور القديمة كان الانسان يفكر بمستقبله، واستشراقه والتنبؤ به. ان التقدم العلمي في العلوم ومنها مثلا الكيمياء وعلوم الحياة والرياضيات وعلوم الحاسبات والعلوم الاخرى تزايد سنويا واصبحت اساسا

لمعظم العلوم الاخرى. فعلى سبيل المثال حدثت تطورات هائلة في الكيمياء وعلوم الحياة كانت بالغة وواضحة فازدهرت فروع جديدة في الكيمياء وشهدت تطورات هائلة في دراسة التركيب الذري والجزيئي كاستخدام اشعة الليزر والاشعة السينية ذات الطاقة العالية والواطئة واشعة كاما واستخدام الصوت والضوء معا في دراسة الذرات والجزيئات والايونات في الحالات الغازية والسائلة والصلبة وحصل تطور بالغ في الكيمياء التحليلية شمل عمليات التشخيص والاستخلاص والفصل وكذلك في الكيمياء الحياتية فقد انبثق منها علم الاحياء الجزيئي والاحيائات الكيميائية Chemical Bionic الذي له شأن كبير في الحاضر والمستقبل، كما حصلت تطورات هائلة في علم الانزيمات اما في علوم الحياة فلم يعد علما واحدا بل اخذت تضم عشرات العلوم الفرعية التخصصية كعلوم التشريح والانسجة والاجنة والخلية ووظائف الاعضاء والوراثة والبيئة والامراض النباتية والحيوانية والبشرية والسلوك والاجتماع والاحافير والتوزيع الجغرافي للحيوان والنبات والتصنيف النباتي والحيواني وتاريخ الاجداد والتاريخ الطبيعي للاحياء وعلوم البكتربا والفايروسات والطفيليات والديدان والحشرات والفطربات والاسماك والطيور واللبائن والهندسة الوراثية والتقانة الاحيائية فقد استخدمت تطبيقات الهندسة الوراثية في المجالات الطبية والزراعية والصناعية واللقاح وانتاج الهرمونات وعلاج الامراض المستعصية التي ليس لها عقار، ففي عام 1997 كان الحدث العلمي المثير الذي اعلن عنه في 27 شباط 1997 وهو ولادة النعجة دوللي بطريقة نقل خلية جسمية متخصصة الى بويضة غير مخصبة لشاة اخرى بعد ازالة نواتها وزرعها في رحم شاة ثالثة والاكثر اهمية لهذا الحدث من الناحية العلمية هو عودة الخلية المتخصصة والمستقرة الى الحالة الجينية بعد فقدانها هذه الصفة. ومع التطورات الاخرى في العلوم الصرفة الاخرى يتضح ان هذا التقدم الهائل كان يخفي وراءه تخلفا في الادراك الانساني لامور اخرى كانتاج كميات كافية من الغذاء لاطعام الانسان في العالم وكذلك المشاكل التي تتعلق بندرة الموارد والطاقة وتزايد مشكلة التلوث والانفجار السكاني.

نظرة مستقبلية للتعليم

سوف يساعد انتشار التعليم في كل دول العالم وثقافاته على بناء المجتمع، فلا مفر من العلاقة الثقافية بين التعليم والمجتمع، وان التعليم ليس الا عنصراً واحداً من عوامل مختلفة ومنها الاجتماعية او الاقتصادية او السياسية او الدينية او العسكرية او الجغرافية، وان الصراع بين القوى العالمية والمصالح المحلية سيكون تحديا تربويا وسياسيا.

سيبقى تحد يتمثل في اهمية مواجهة الصراعات الدينية والعرقية والاثنية والثقافية والقومية وان قدرة التعليم المنظم في المدارس والجامعات على التقليل من هذه الصراعات محكومة الى حد بعيد بالادارة السياسية للقياديين في المجتمع نحو مشاركة جدية وايجابية للتربية والتربويين في مثل هذه القضايا، غير انه لا يمكن الحد من مستويات هذه الصرعات من دون مشاركة تربوية ولن تكون الاستراتيجيات العسكرية كافية بحد ذاتها.

ويجب ان ينظر الى المعلمين بدءا من مرحلة ما قبل المدرسة وصاعدا على انهم وسيلة للتفاهم الثقافي والديمقراطي، فان ذلك يعني ديمقراطية معقمة وهذا يعتمد بدوره على تطور التفاهم بين الثقافات او التفاهم متعدد الثقافات.

من التحديات ذات الصلة الماثلة امامنا ان نصل الى طرق واقعية لسد الفجوة الهائلة بين الاغنياء والفقراء في العالم، هذه الفجوة موجودة في المدن ذاتها، او بين المدن والمناطق الريفية او في الدول ذاتها او بين دولة واخرى وفي مناطق العالم المختلفة وبينها وهو الحد الفاصل بين الشمال والجنوب.

ان التدخلات التربوية لها اهمية محورية عند مجابهة هذا التحدي، كما يجب ان يزود التعليم الدارسين بالمهارت الاكاديمية والتقنية الضرورية للعديد من الدول النامية التي تسعى الى الابتعاد عن الزراعة كمورد رزق وحيد ويمكن للتعليم الراسخ في مفهوم المشاركة ان يساعد على توجيه المواطنين لاتخاذ قرارات صائبة بشأن الانفاق والتوفير، والاستثمار في التعليم والرعاية الصحية.

من اجل التصدي لهذه التحديات وغيرها يتعين على مجتمعات العالم ان تعيد احياء مؤسساتها وحكوماتها، كما يتعين عليها ان تعزز الحياة المجتمعية ولعل التعليم احد طرق

تحقيق ذلك اما المشاركة فيجب ان تكون موضوعا تربويا رئيسيا على المستويات المختلفة كافة. لكن الخطوة المهمة الاولى التي يتعين على الحكومات ان تتخذها هي ان توفر الاموال والمرافق الضرورية لنظام التعليم العام الذي يبدأ بالصغار في مرحلة مبكرة من اعمارهم ويمتد الى الكبار. اما نظام التعليم العام الخاضع لتوسعات واصلاحات فيجب ان يكون محوره مدارس ابتدائية وثانوية فعالة.

ومن اجل مجابهة التحديات لابد ان يكون للتعليم في كل مستوياته تاثير كبير في الطلبة، أي يجب ان يكون التعليم فعالا، والتعليم الفعال الذي يمكن مجتمعا ما من مجابهة التحديات المقبلة يجب ان يكون على ثلاث قضايا رئيسية. من الاهداف المجتمعية المهمة ان تخلق نظاما تعليميا فعالا للطلبة كافة على ان يشمل ذلك الطلبة الاقل حظا من الناحية الاقتصادية وان تساعد الطلبة كافة على تحقيق مستويات عليا من التحصيل الاكاديمي.

يتعين على التربويين في المستويات المختلفة ان يضعوا خطة للمناهج والبرامج يمكن ان تراعي افضل ما يمكن مراعاته من الفروق الفردية في اهتماماتهم ومواهبهم وحوافزهم وثقافاتهم على التربويين ان يدركوا ان الانجازات الاكاديمية والانشطة الفكرية في التعليم لا يمكن وضعها باي شكل من الاشكال عن التطورات الاجتماعية والعاطفية والاخلاقية.

على قادة المجتمع جعل التعليم يساهم مساهمة فعلية في مواجهة التحديات ومنها خلق مجتمعات تعلم لان المدرسة وحدها ليست كافية مع دور رئيس للمدارس العامة واستحداث مجتمعات التعلم تتيح التعليم لجميع السكان وبناء مجتمعات اكثر تماسكا مع شبكة للجمعيات المدنية مع مدارس مجتمعية.

ويتطلب التعليم مستقبلا :

- مناهج جديدة متكاملة مع الوسائط الفعالة متعددة التفاعلية.

- وسائط متعددة التفاعلية يعكف على اعدادها علماء دوليون بارزون.

- مستويات الاتصالات وتقنية الحاسوب الملائمة لمستوى كل طالب لتنشيط الابداع والابحاث.

440

- تغيير الكتب المدرسية ومجموعة واسعة من البرمجيات الدراسية واجهزة الحاسوب الشخصي والاقراص المبرمجة والتلفاز التربوي.

- ادوار جديدة للمعلمين وتدريب جديد اثناء الخدمة وخارجها لجمع المعرفة.

- مشاركة قوية بين المنزل والمدرسة.

- طريقة جديدة لتقويم وتحديد قدراتهم وميولهم.

- تنوع التعليم بعيدا عن الاشكال التقليدية.

اما مدارس المستقبل فتتطلب توازن بين اهدافها العامة ومن هذه المتطلبات:

- تجهيز الطلاب لمستقبل منتج.

- رعاية القدرات الفكرية لدى طلابها.

- تدريبهم على فهم ثقافتهم.

وتقوم هذه المدارس بـ:

- ان تقرر اولوياتها وبرامجها فيما يتعلق بالمناهج ضمن اطار عام وقوي للمناهج والمقاييس.

- ان تقرر اولوياتها فيما يتعلق بالانفاق.

- القيام بعمليات تقويم الطلاب الخاصة بها في المستويات كافة واعداد تقارير سنوية بالنتائج.

تضم مدارس المستقبل خمسة عناصر ثبت دوليا انها حاسمة بالنسبة الى التعليم المدرسي الناجح :

- تركيز نشط على كل مدرسة على حدة.

- المرونة في الاستمابة لاحتياجات طلاب المدرسة ذاتها.

- التزام المجتمع المحلي بنظام التعليم المدرسي.

- معايير واقعية.

- المساءلة امام المجتمع المحلي.

الدراسات المستقبلية في العلوم

ان معظم الدراسات المستقبلية الحالية التي انتهت بعام 2000م والبعض القليل عند حتى عام 2025م ويتطلب لاجراء مثل هذه الدراسات:

- وسائل برامج خاصة لاجراء عملية التنبؤ.

- تحتاج الى العديد من الخبراء والفنيين والمبرمجين ليسوا من اختصاص واحد ففيهم علماء الرياضيات والكيمياء والفيزياء وعلوم الحياة وغيرهم وقد اطلق عليهم مهندسو المعرفة (مجمع الفكر0 او خزان الفكر think tank.

وفي ضوء ذلك من الممكن ان نقسم الدراسات المستقبلية في العلوم الى ثلاثة انواع وهذا التقسيم اعتباطي سنستعمله للتبسيط فقط، اذ ان من الصعب فصل الانواع الثلاثة من الدراسات الواحدة عن الاخرى وهذه الانواع:

- الدراسات التي تعتمد على التنبؤ بما سيحدث في مجال علمي معين مثل التطور الذي سيحدث في مجال الهندسة الوراثية والحاسبات والكيمياء والرياضيات والفيزياء.

- الدراسات المستقبلية الشاملة التي تعتمد على الحدس والتي تنظر الى تأثير الانجازات العملية الحالية على مستقبل الانسانية سواء على صعيد واحد او اكثر.

- الدراسات المستقبلية الشاملة التي تعتمد على معلومات احصائية مفصلة ضمن برامج رياضية او نماذج حاسوبية وتستعمل في مجالات متخصصة مثل التنبؤ باستهلاك الطاقة.

نظرة مستقبلية للتعليم

سوف يساعد انتشار التعليم في كل دول العالم وثقافاته على بناء المجتمع، فلا مفر من العلاقة الثقافية بين التعليم والمجتمع، وان التعليم ليس الا عنصراً واحداً من عوامل مختلفة

ومنها الاجتماعية او الاقتصادية او السياسية او الدينية او العسكرية او الجغرافية، وان الصراع بين القوى العالمية والمصالح المحلية سيكون تحديا تربويا وسياسيا.

سيبقى تحد يتمثل في اهمية مواجهة الصراعات الدينية والعرقية والاثنية والثقافية والقومية وان قدرة التعليم المنظم في المدارس والجامعات على التقليل من هذه الصراعات محكومة الى حد بعيد بالادارة السياسية للقياديين في المجتمع نحو مشاركة جدية وايجابية للتربية والتربويين في مثل هذه القضايا، غير انه لا يمكن الحد من مستويات هذه الصرعات من دون مشاركة تربوية ولن تكون الاستراتيجيات العسكرية كافية بحد ذاتها.

ويجب ان ينظر الى المعلمين بدءا من مرحلة ما قبل المدرسة وصاعدا على انهم وسيلة للتفاهم الثقافي والديمقراطي، فان ذلك يعني ديمقراطية معقمة وهذا يعتمد بدوره على تطور التفاهم بين الثقافات او التفاهم متعدد الثقافات.

من التحديات ذات الصلة الماثلة امامنا ان نصل الى طرق واقعية لسد الفجوة الهائلة بين الاغنياء والفقراء في العالم، هذه الفجوة موجودة في المدن ذاتها، او بين المدن والمناطق الريفية او في الدول ذاتها او بين دولة واخرى وفي مناطق العالم المختلفة وبينها وهو الحد الفاصل بين الشمال والجنوب.

ان التدخلات التربوية لها اهمية محورية عند مجابهة هذا التحدي، كما يجب ان يزود التعليم الدارسين بالمهارت الاكاديمية والتقنية الضرورية للعديد من الدول النامية التي تسعى الى الابتعاد عن الزراعة كمورد رزق وحيد ويمكن للتعليم الراسخ في مفهوم المشاركة ان يساعد على توجيه المواطنين لاتخاذ قرارات صائبة بشأن الانفاق والتوفير، والاستثمار في التعليم والرعاية الصحية.

من اجل التصدي لهذه التحديات وغيرها يتعين على مجتمعات العالم ان تعيد احياء مؤسساتها وحكوماتها، كما يتعين عليها ان تعزز الحياة المجتمعية ولعل التعليم احد طرق تحقيق ذلك اما المشاركة فيجب ان تكون موضوعا تربويا رئيسيا على المستويات المختلفة كافة. لكن الخطوة المهمة الاولى التي يتعين على الحكومات ان تتخذها هي ان توفر الاموال

والمرافق الضرورية لنظام التعليم العام الذي يبدأ بالصغار في مرحلة مبكرة من اعمارهم ويمتد الى الكبار. اما نظام التعليم العام الخاضع لتوسعات واصلاحات فيجب ان يكون محوره مدارس ابتدائية وثانوية فعالة.

ومن اجل مجابهة التحديات لابد ان يكون للتعليم في كل مستوياته تاثير كبير في الطلبة، أي يجب ان يكون التعليم فعالا، والتعليم الفعال الذي يمكن مجتمعا ما من مجابهة التحديات المقبلة يجب ان يكون على ثلاث قضايا رئيسية. من الاهداف المجتمعية المهمة ان تخلق نظاما تعليميا فعالا للطلبة كافة على ان يشمل ذلك الطلبة الاقل حظا من الناحية الاقتصادية وان تساعد الطلبة كافة على تحقيق مستويات عليا من التحصيل الاكاديمي.

يتعين على التربويين في المستويات المختلفة ان يضعوا خطة للمناهج والبرامج يمكن ان تراعي افضل ما يمكن مراعاته من الفروق الفردية في اهتماماتهم ومواهبهم وحوافزهم وثقافاتهم على التربويين ان يدركوا ان الانجازات الاكاديمية والانشطة الفكرية في التعليم لا يمكن وضعها باي شكل من الاشكال عن التطورات الاجتماعية والعاطفية والاخلاقية.

على قادة المجتمع جعل التعليم يساهم مساهمة فعلية في مواجهة التحديات ومنها خلق مجتمعات تعلم لان المدرسة وحدها ليست كافية مع دور رئيس للمدارس العامة واستحداث مجتمعات التعلم تتيح التعليم لجميع السكان وبناء مجتمعات اكثر تماسكا مع شبكة للجمعيات المدنية مع مدارس مجتمعية.

ويتطلب التعليم مستقبلا :

- مناهج جديدة متكاملة مع الوسائط الفعالة متعددة التفاعلية.

- وسائط متعددة التفاعلية يعكف على اعدادها علماء دوليون بارزون.

- مستويات الاتصالات وتقنية الحاسوب الملائمة لمستوى كل طالب لتنشيط الابداع والابحاث.

- تغيير الكتب المدرسية ومجموعة واسعة من البرمجيات الدراسية واجهزة الحاسوب الشخصي والاقراص المبرمجة والتلفاز التربوي.

- ادوار جديدة للمعلمين وتدريب جديد اثناء الخدمة وخارجها لجمع المعرفة.

- مشاركة قوية بين المنزل والمدرسة.

- طريقة جديدة لتقويم وتحديد قدراتهم وميولهم.

- تنوع التعليم بعيدا عن الاشكال التقليدية.

اما مدارس المستقبل فتتطلب توازن بين اهدافها العامة ومن هذه المتطلبات:

- تجهيز الطلاب لمستقبل منتج.

- رعاية القدرات الفكرية لدى طلابها.

- تدريبهم على فهم ثقافتهم.

وتقوم هذه المدارس بـ:

- ان تقرر اولوياتها وبرامجها فيما يتعلق بالمناهج ضمن اطار عام وقوي للمناهج والمقاييس.

- ان تقرر اولوياتها فيما يتعلق بالانفاق.

- القيام بعمليات تقويم الطلاب الخاصة بها في المستويات كافة واعداد تقارير سنوية بالنتائج.

تضم مدارس المستقبل خمسة عناصر ثبت دوليا انها حاسمة بالنسبة الى التعليم المدرسي الناجح :

- تركيز نشط على كل مدرسة على حدة.

- المرونة في الاستمابة لاحتياجات طلاب المدرسة ذاتها.

- التزام المجتمع المحلي بنظام التعليم المدرسي.

- معايير واقعية.

- المساءلة امام المجتمع المحلي.

الدراسات المستقبلية في العلوم

ان معظم الدراسات المستقبلية الحالية التي انتهت بعام 2000م والبعض القليل عند حتى عام 2025م ويتطلب لاجراء مثل هذه الدراسات:

- وسائل برامج خاصة لاجراء عملية التنبؤ.

- تحتاج الى العديد من الخبراء والفنيين والمبرمجين ليسوا من اختصاص واحد ففيهم علماء الرياضيات والكيمياء والفيزياء وعلوم الحياة وغيرهم وقد اطلق عليهم مهندسو المعرفة (مجمع الفكر0 او خزان الفكر think tank.

وفي ضوء ذلك من الممكن ان نقسم الدراسات المستقبلية في العلوم الى ثلاثة انواع وهذا التقسيم اعتباطي سنستعمله للتبسيط فقط، اذ ان من الصعب فصل الانواع الثلاثة من الدراسات الواحدة عن الاخرى وهذه الانواع:

- الدراسات التي تعتمد على التنبؤ بما سيحدث في مجال علمي معين مثل التطور الذي سيحدث في مجال الهندسة الوراثية والحاسبات والكيمياء والرياضيات والفيزياء.

- الدراسات المستقبلية الشاملة التي تعتمد على الحدس والتي تنظر الى تأثير الانجازات العملية الحالية على مستقبل الانسانية سواء على صعيد واحد او اكثر.

- الدراسات المستقبلية الشاملة التي تعتمد على معلومات احصائية مفصلة ضمن برامج رياضية او نماذج حاسوبية وتستعمل في مجالات متخصصة مثل التنبؤ باستهلاك الطاقة.

References

- Robert M., Gane, Essentials of learning for instruction, 1975, Holt Reinehort and Winston.
- Travers, essential of learning, 1997, N.Y. Macmman.

- Borich, Gary D.: Effective teaching methods. Second edition, 1992, Macmilan.

- Vaizey, J., and Sheehan, J., Resources for education, 1968, Allen and UNWIN.

- Andrew Taylor, Frances Hill "Quality management in education" in (organization effectiveness and improvement in education), edited by Alma Harris 1997, open university, press, Backing ham.

- Jerry Banks: Principles of quality control, 1989, John Wiley: sons, N.Y.

- AECT, Educational Technology: Definition and Glossary of Terms, Vol. 1, 1977, Vol. 1. Washington, DC: AECT.

- Mackenzie, Norman and others: open learning systems and problems in post- secondary education, 1975, Paris: Unesro press.

- Lenrer. J (2000): learning disabilities theories diagnosis and teaching diagnosis (6th ed) (2000), Houghton Miffin Co. Boston.

- Mercer, G.: students with learning disabilities (1997), Columbus OH prentice- Hall INC.

- Polloway, EA: Patton. J.: strategies for teaching learners with special. (1993), Merrill – Newjersy, Columbus- Ohio.

- Boyd. Gary: the shapping of educational technology by cultural politics and vice versa. Educational and training technology international (1991), 28(2).

- American Association for advancement of science, science for all Americans (1989), project 2661, Washington, D.C.

- Anthony D.F. and Dean L.C. science for all children elementary school methods, (1998) MSA, Waveland press.

- Croline, Mc. G. science: technology and science (1999). Handbook, the association for science education.

- Chun, S., Scientific literacy an educational goal of the past two costing NARST Annual Meeting (1999), Boston, Massachusetts.

- Korla, R.M.: popularizing science in schools (2000), Delhi. Ram Prmto-graph.

- Encyclopedia Americana, Vol, 15, Japan: (1981), education.

- Carl Parsons Quality improvement in education, (1994), David Fulton Publisher Ltd- London.

- John Dewy, philosophy of education (1985) Littlefield Adams and Co. N.Y.
- Correa H: A national Methods in educational Planning and Administration, (1979), N.Y. David Mcky.
- Bozeman, W.C. Educational Technology (1995) best practices from American school, N.Y. Eye on education.
- Ellington, Henry, and others, Handbook of educational technology (1995), Kongan page, 3rd edition.
- Gagne, R.,: The conditions of learning (1985), New York, Holt, Rinehart and Winston.
- Garadner, H., Creating minds, (1993), New York.